GARDEN
GARDEN
GARDEN
GARDEN

Hold On To Your Hat

Interviews with Bern Porter

edited by mIEKAL aND

2012
Xexoxial Editions
West Lima, Wisconsin

© 2012 Mark Melnicove for the Estate of Bern Porter

© 2012 Photography by Steve Random
© 2012 Photography by Amy Hufnagel
© 2012 Photography by Joel Lipman
© 2012 Photography by Reed Brugger

Back cover photograph by Amy Hufnagel

All interviews printed or reprinted with permission of the authors.

"Physics for Tomorrow" reprinted and edited in 1999 as "Physics Today" (Roger Jackson, Publisher)

Hat-isms courtesey Fred Belinsky, Village Hat Shop (www.VillageHatShop.com)

Much gratitude to Ben Meyers, Justin Katko & Monty Cantsin—Xexoxial interns extraordinaire—who helped compile, shape and design this collection.

ISBN-10: 1-936687-07-0
ISBN-13: 978-1-936687-07-7

published by

Xexoxial Editions
10375 County Hway Alphabet
La Farge, WI 54639

www.xexoxial.org
perspicacity@xexoxial.org

Table of Contents

mIEKAL aND
Interview with Bern Porter — 2

Steve Random
Photographs — 62

mIEKAL aND, Elizabeth Was and Ben Meyers
Questions for Bern Re: The Institute of Advanced Thinking — 74

Joel Lipman
Photographs — 85

Dick Higgins
Interview with Bern Porter — 93

Amy Hufnagel
Photographs — 125

Mark Melnicove
Interview with Bern Porter — 133

Read Brugger
Photographs — 146

Phobrek Hei and Sasha K.
Interview with Bern Porter — 153

Seth Tisue and Brad Russell
Interview by Mail with Bern Porter — 161

Judith Hoffberg
Interview with Bern Porter									172

Bern Porter
Physics for Tomorrow									176

Bern Porter
Introduction to *Xerolage 16*							186

Bern Porter
Significant Content									188

Bern Porter
The Preposition Song									190

Amy Hufnagel
See(MAN)TIC / Bern Porter								191

Andrew Russ
Books by and about Bern Porter							195

Andrew Russ
Review of Bern Porter's *I've Left*							199

Hat-isms											208

Bibliography										216

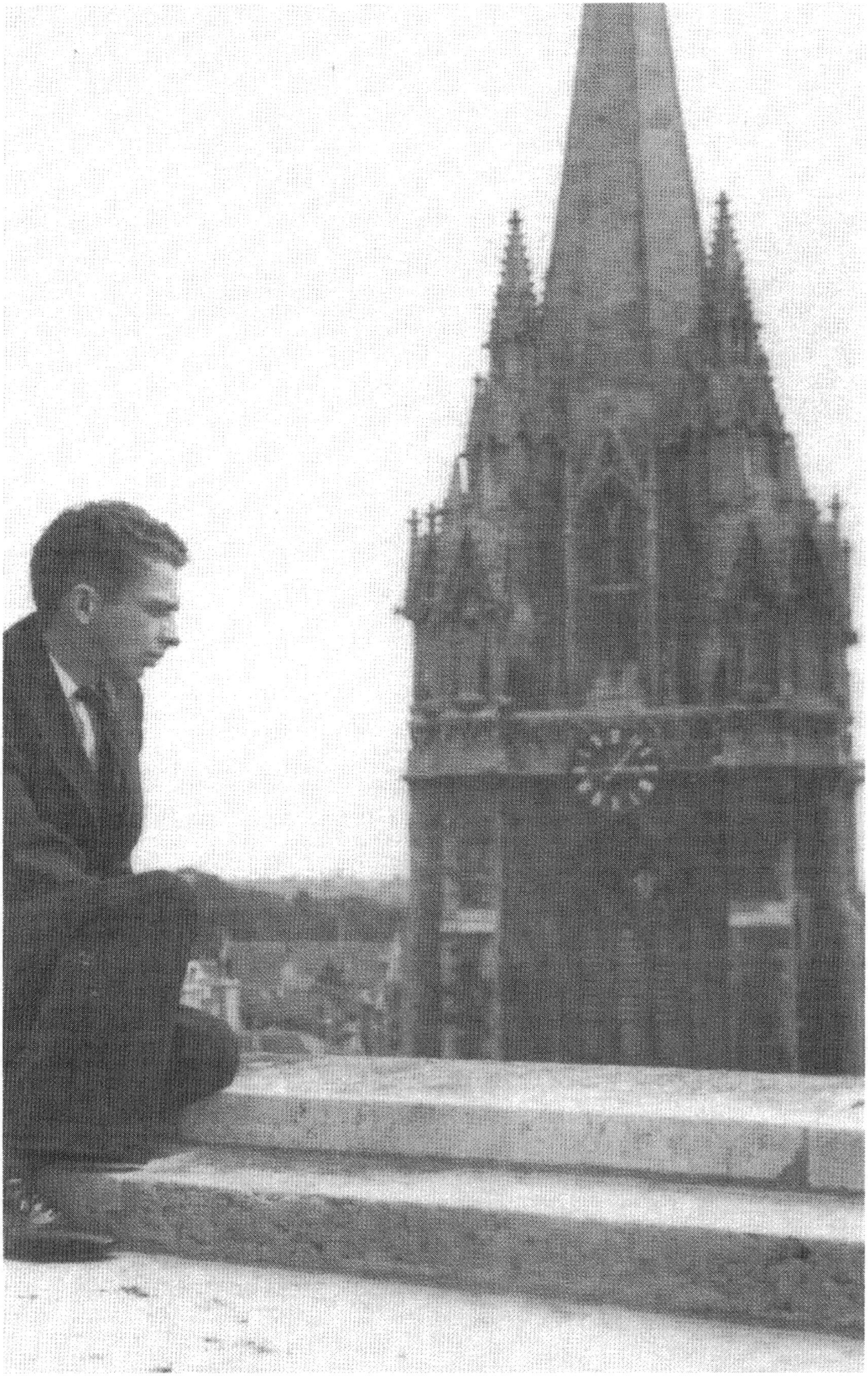

Bern Porter interviewed by mIEKAL aND
November 1989

Bern Porter: 1341 Williamson Street, Madison, Wisconsin in November 1989, and we're talking with mIEKAL aND about the Institute of Advanced Thinking in Belfast, Maine. We have already reproduced and copied three ten-year-old tapes. We've already made thirty minutes on the tape to be called *Williamson Street Nights*. We are in the process of preparing a manuscript, a typescript for a movie. We come now to the Institute itself, and the first point is one of historical importance, I feel.

mIEKAL aND: *How did the original idea for the Institute come about?*

BP: Historically we find that in Greece, in Athens, in Rome, in the time of the Romans and Greeks that they were able, in their sandals and togas in public squares, to sit around and talk about, amazingly, the same things we talk about. They were able to conjecture atoms, molecules, interactions of interplay, and they developed what we call hypotheses, that is, of chord structures, of reason, of logic, and put together their understanding of how nature works, and thus gave birth, so to speak, of what we today call physics. Physics is actually the study of nature, the duplication of nature, of what nature does.

So here, in those early days before Christ, were men discussing things that we discuss today. Caesar and his armies passed over the Alps and went north into what is now France, was called then Gaul; went on into what is now England and what is now Scotland and Ireland, where they brought their ideas and in conflict fused them with the Celts. The Celts were a mystic religious group who had ideas which interpreted nature for them.

And finally around the year 1,000 people from what is now Scotland and Ireland came to the coast of Maine. Arriving at the coast of Maine around the year 1,000 they were quite amazed to find that some 50,000 people, native peoples of Maine had lived there some 10,000 years. And during these 10,000 years they developed an incredible array of things which we use today. And I think it worthwhile just to go down through a list of some of things which were in existence as a result of 10,000 years of living along the coast of Maine, and which were discovered by the white man around the year 1,000. For example, the native Indians—we don't call them Indians, I'm sorry, we call them native peoples—invented and were using what we today call popcorn.

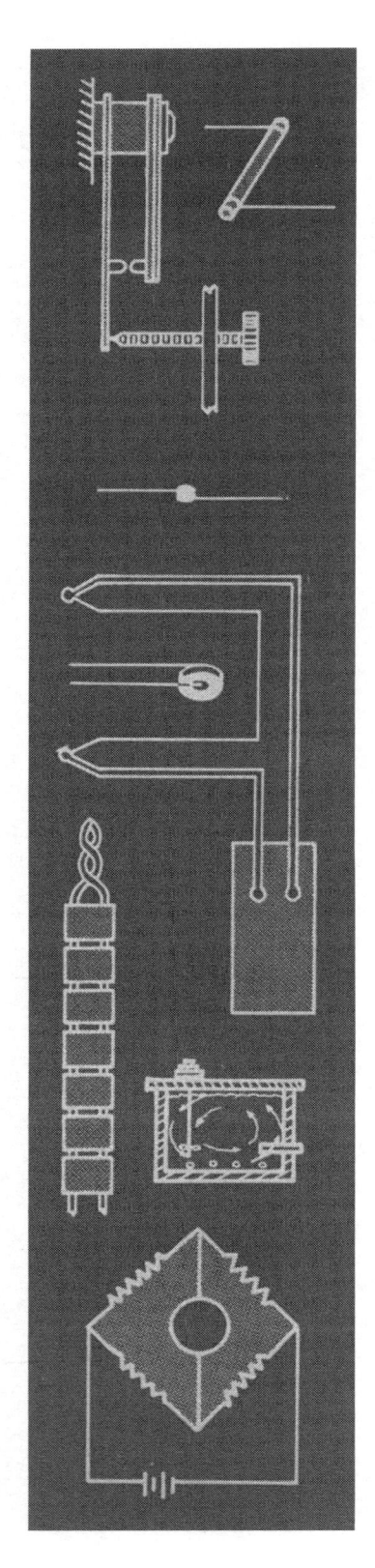

They also invented and pioneered and used daily in their lives corn flakes, flapjacks, and they had maple syrup. They had drugs like quinine and cocaine and cascara sagrada, witch hazel, oil of wintergreen, petroleum jelly, and they made dyes. This is a culture who over a period of 10,000 years developed a great many things, and among them, of course, is what we have today come to call a Bill of Rights, a Declaration of Independence.

There were many plants living and growing and from these medicines were made. Such medicines as water heron, many such plants like goldenrod and ginseng and balsam, butternut and black cherry, ladyslipper and moosewood. And all of these were converted by these people, these native people, into medicines. And of course from the plants they made fillings for mattresses, they made pillows, they made dyes, they made beverages. And while we refer to them as *Indians* and backward and whatever, they really developed to a very, very high degree many of the things that we have. From animal bones of course they made sewing thread, bow strings, pendants, rattles, needles, pipes, dice, bags and containers. They went on to make tools, ornaments, fuel, fertilizer, food and clothes. They sustained themselves from the land. All of this before the year 1,000.

The native peoples coming from Scotland and Ireland, the white people, of course brought disease, they brought the Bible, they brought guns, and what was originally 50,000 people is now reduced to something like 3,000, divided off into about 5 tribes, or what they call nations, that are now in Maine. The white people came, settled along the coast from 1240 to 1640, bringing these ideas originally from Rome and Greece and then the Celts, and then fused with the native ideas. So that by the time *my family*, the Porter family, came from Ireland and Scotland around 1830, the ideas along the coast had been pretty well established in Popham and Casteen, but were brought to some kind of fusion and reorganization in 1830 in what was Porter Settlement.

Porter Settlement is one of the smallest subdivisions in Maine—it has no post office and no store, and only in the last five years has it been indicated on the maps of Maine—about three and a half miles outside of the town of Houlton. My ancestors from 1830 on started in a log cabin, cleared the land, and eventually moved from the log cabin into a house, a one-story house which they later developed into two stories with the additional barns, a typical Maine farm. By the time I was born in 1911 there were twenty-three Porters, and with them they had all of the ideas which we now use in the Institute. So this has come down to me through my family, 1830 down to me in 1911, and has been functioning in Tehunseh, Maine for the last seventy-five years. I claim to have invented what we call Mail Art today, in Porter Settlement, Maine. Marcel Duchamp, independently of me, in France, was also instituting his conception of Mail Art, and also of *founds*.

So we find this interwoven interrelation dating back to the Greeks, Celts, the native peoples, and come down to 1989 where I would like to come to the basic question. The name is the Institute of Advanced Thinking. Journalists come along and they say, *Well, this is a think tank. And if it's a think tank then how many people are involved?* Well it happens that there are thirty-one of us. We're spread throughout the world—there's one in Japan, there's one in New Zealand, there's one in Germany, West Germany, Italy—spread around the world and Belfast, Maine is the headquarters, the world headquarters. And over the years some have come to Belfast, and I in turn visit them, by means of ships, flights of QE II into Rotterdam. I advise them I'm coming, they meet me, I meet them in their home environment. It's interesting that all thirty-one of us are freelance. That is to say, we're not connected with any university, we're not connected with any government, and what we conceive of, we do not patent or copyright. We give free to the world and to all.

Well the question comes up....

Interviews with Bern Porter

MA: *What does the Institute think about on Tuesdays?*

BP: Well, the Institute used to call Tuesdays *World Problem Days*. Now we simply call them days where we do indeed have up for consideration an array of problems which really have all the characteristics of haze, so real and different are they. So we have *haze days*, you might say.

And we're considering what? Well, we're considering political and religious differences, disease, war, moral crises, population explosion, uncontrolled technology, pollution. And on top of this we have obvious things like harnessable ocean tides, harnessable winds, harnessable sun, harnessable alcohol-producing vegetation. And of course it's obvious, I guess, that none of these will probably be harnessed to their fullest potential—that is, tides, winds and sun.

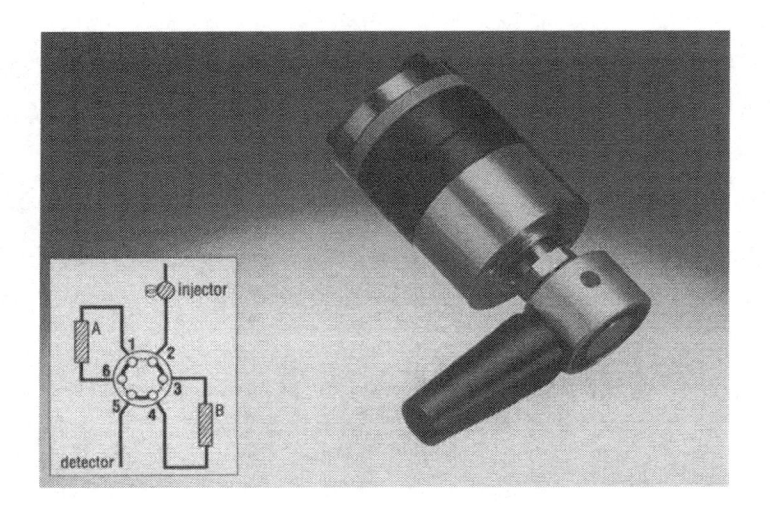

And obviously all this is only a *surface dressing*, about which everyone already knows, or has heard about. For example there still are disappearing animals, birds, botanical specimens; there's arms production, sale and distribution; there's consumption and waste of natural resources; there's the folly of nuclear power; there's terrorism; there's corruption; there's the setting up and collapse of governments; there's the starving and there's the homeless, all of which you've heard about.

Then on top of that we have of course the latest topic, which concerns exploding stars, and the passage of our present life-giving sun to a cold black cinder, while our lifeless Earth goes into an eternal deep freeze. Or as we might say, why, pray tell, is our oxygen different now than it was eighty million years ago?

HOLD ON TO YOUR HAT

MA: *It's such a mix, there is no distinction over what is needed and what is just happening!*

BP: Well, that is correct. There is no distinction between what is needed and what is happening, because we're in a state of flux. Now for example, on the side of betterment: our group of thirty-one—myself, thirty-two—we might start off with education, which in our time is really reduced to a license to work, and how expensive and wasteful it is, and has currently become! Say it's up to $16,000 a year for one college student as opposed to the Institute's Three Major Solutions, at least for the early years: We give thought to this, it's called *SciEd–Sci* standing for science, which is really physics—so we're combining physics and education. And the first of these, which were quite often ridiculed but are quite possible: the body of a pregnant mother is fitted with a stomach electrode band, through which is introduced for eight months by electronic means to the developing mind of the growing fetus, all the instructional requirements of three languages, mathematical essentials, reading, history, philosophy, drama, poetics; and such amounts of forms that at birth the child is already the equivalent of a Ph.D. We have advocated this and are waiting for it, frankly, to be followed up. That we offer as Number 1, as a substitute for our present education.

 Number 2: We have elaborate busing, buildings, facilities and extracurricular activities like sports, dramatics, classrooms, gymnasiums, dining area, and playgrounds. All of these are simply to be eliminated by having all students *stay at home*, to study in cubicles, to study in their private rooms, connected by television to a Learning Clinic in the center of each state. And for three hours a day, seven days a week, each student is instructed via television in three languages, essential mathematics, reading, philosophy, religion, literature, history and science.

 Periodic reviews of materials covered are sent to students by mail, along with necessary examinations, returnable to the Clinic for grading and credit. This is a form which we call *SciEd*. This is the second suggestion.

 Now, Number 3: Having gone through either of these two, or all of these foregoing educational techniques for the early years, the students are then ready for apprenticeships in the field of their choice. Apprenticeships have gone out of style, it's something that was done in the early days—a father passed on his talent to his son. A type of training practice that was favored and common in this country's development stages and now lies bypassed, unused and forgotten, as I've said.

The student is also ready for the specialized schooling at particular professional training centers, like a veterinarian, medical doctor, lawyer, engineer. Centers devoid of the distractive elements of the present universities, like football, track, rowing, dramatics, fraternities, campus newspapers, dances and social affairs.

So we have these three suggestions.

MA: *General education seems to have gone.*

BP: Well, not only has it gone, but the BA degree, that catch-all diversion, is now to be patched up within employable specifics, while the methods we have noted fit almost immediately into an income return. Perhaps more significant than monetary results is the recognition through achievement, as, for example, Anaïs Nin, who early on had formal assistance, read everything in sight, chose a master to work with and for, and then took off—not only to produce, but to affect, influence and be someone.

MA: *How does the Institute fit into all this?*

BP: Well, I've given the history from the earliest times down to 1989, where there are thirty-two of us—free of governments, free of academics—who consider all the problems I've listed, but are concerned with, as I've said, the three forms of education that I've just noted.

MA: *Is advanced thinking different from stream of consciousness association or altered states?*

BP: Well, the word *advanced*, as the Institute employs it, means a rejection, a complete rejection of all past methodologies, with an obvious forward progression in a manner that the current press technology calls *breakthroughs*. We're interested in producing what are called *breakthroughs*. But at the Institute, a breakthrough is considered anything that's obvious, rational and reasonable, as advanced through the greatest possible number of people. Among the free systems of education already cited—learning in the womb, learning at home, study apprenticeships—the first two of these, by modern electronic means, are fine examples.

I would like to give some other examples here, but before I do so, I would like to point out that people ask me, *What do you do when you are doing advanced thinking?*

And the answer is, we are making the complex simple, that whatever is reasonable and obvious we are undertaking. We have developed out of this what is called the Bern Porter Theory of Simplistics, which can be reduced more or less as follows: *It is easier to complicate things than to simplify them.* This is a man-made truism. Rather than make things simple or obvious or reasonable, we tend to mess them up and complicate them. Whatever is reasonable we make unreasonable, whatever is obvious we obscure, whatever is easy we make it difficult, whatever is difficult we never make easy, and whatever is complicated we tend to make more complicated and whatever is complicated is never made easy. So whatever is obvious is never done, whatever is reasonable is never done.

These are the elements of Simplistics.

So when you ask me what is advanced thinking, it reduces to making the complicated simple. A very staggering example that's going on right now in Europe—in connection with this view of looking at old things in a new way—there are countries in Europe who've looked at Marxism for forty years, who've looked at Leninism for forty years, who are now looking at old things in a new way. People for forty years have been bound up with the theories of Marx, the theories of Lenin, the theories of Stalin. And in physics we're looking at the ideas of Newton and Einstein. Everything is now under re-examination. These people in Europe after forty years of looking at this old thing in a new way, will suddenly look at socialism, they will throw out communism, they will look at democracy, and they will add to, and build upon, and build a greater democracy, as a result of looking at an old thing in a new way. And this is a part of the Theory of Simplistics that we advocate and develop at the Institute. Please, please, look at old things, old attitudes, common things around you, and look at them in a new and fresh way.

Now a very classic example of this is our friend and honorary member of the Institute, Buckminster Fuller. He lived on an island, his family lived on an island off the coast of Maine, for about eighty years. He was schooled by ships going by, he would row out to the ships who were taking out lumber and bringing in paper and salt and products that were used in Maine.

He looked at the globe, a sphere, a globe of the Earth, twelve inches in diameter, he looked at it for years, and said, *I'm tired of looking at this thing. What can I do to it?* Well, it turned out he said, *I can cut it up.* And it turned out that since it was a sphere and a globe, a model of the Earth, each section he cut up had to be a triangle. Having cut it up into triangles, he laid the triangles on the floor flat and looked at them in a fresh and new way, and to his amazement he could understand the meaning of the equator, the tropic of Cancer, the tropic of Capricorn, the North Pole, the South Pole, the relation of the lands to the water; he could see the world and the continents in a new way. He'd been hampered all his life by a globe, now suddenly the globe was reduced to pieces and the pieces were flat on the floor, a flat surface, two-dimensional. He was able then to see in a fresh way. This is for us a very classic example of looking at old things in a new way.

The people of Europe are now looking at old systems of communism, Marxism and the rest, and are seeking a new way, what we call a breakthrough.

Another very interesting example is a solution to problems—take a simple thing like gun laws, or guns, handguns. It is said nationwide, worldwide, that handguns are a problem. Well, if they're a problem then isn't it obvious that you attack them by first not permitting their manufacture anywhere in the world? Stop their manufacture. Then find people who own them, actually go door to door and *remove* them. Either their owners pay a fine or give up their handgun; and prevent, stop the sale. Just remove handguns from all markets, from all use, and make them disappear, just decree that they are a problem.

Well, obviously this will never be done. Just the same as the school system that I have advised, and the womb studies, probably will never come about. This is typical of man in his continual desire, innate desire to make messes. To make complicated what is already complicated, and never, never to simplify.

So, there's the case of the handgun, I've mentioned the globe, and now we might go on to a thing that's called a forty-hour week. We have unemployment. We have the trouble of what to do with our leisure time, and how to make a livelihood, and the problem is what do we do about this. Well, the problem is the forty-hour week, the five-day workweek, holidays, weekends. In short our present entire system of employment is eliminated. The twenty-four hours of each day and the seven-day week are divided into three shifts of eight hours each. No one works more than thirty-two hours a week, that is, four eight-hour periods, in a shift period of his choice, while production proceeds continuously, full time, for 168 hours a week.

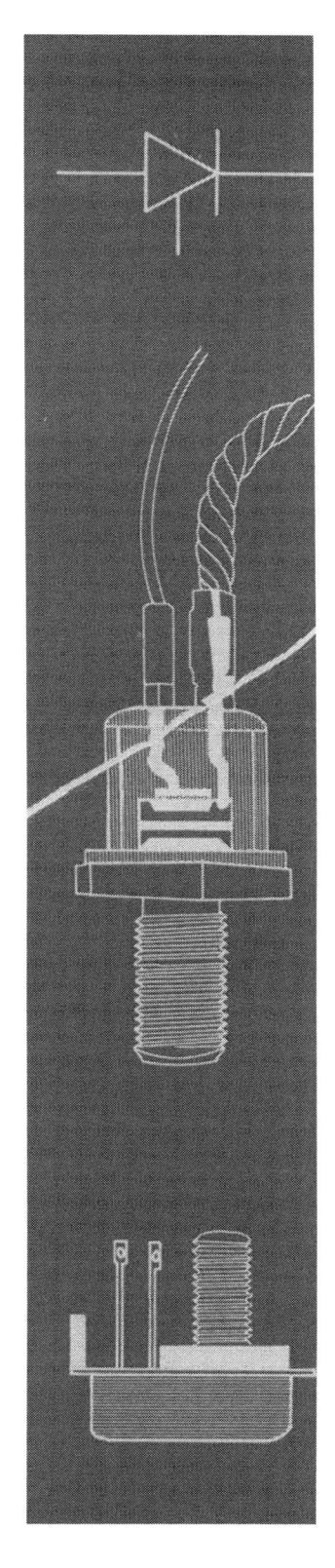

When I was considering this, a seventy-two-story skyscraper was being built across the street. Under the present system, the site is literally crawling with workers for forty hours, then stops dead deserted for 128 hours. A total standstill, with nothing else happening except a guard on watch, patrolling. Now under the plan of the Institute, the work would proceed 168 hours a week, many workers would have employment, and the building would be finished in one-fourth the time. And as a further effect, the present pool of unemployed would be absorbed. Apprentices would be training to fill in this and future needs, with every able-bodied person having a paying job for thirty-two hours, but without overtime. The resulting leisure hours would be obvious for each person, opening a whole new range of possibilities. Where some critics would suggest that crime, alcohol, drugs, auto accidents, thefts would increase, the reverse would be nearly true, where suddenly people would be finding time to do what they truly always wanted to do, but never did, because they *had to go to work*. Hobbies would thrive, for the healthy betterment of everyone. It's interesting that in connection with this time-rearrangement, one's personal twenty-four-hour day would best be divided into four six-hour segments for easy planning and survival, that is, from six in the morning to noon, from noon to six at night, from six at night until midnight, and from midnight until six AM. And I think I've mentioned that wherever this proposal would be carried out, the greatest number of people would benefit. And this idea, as we offer it to the world, is that we dissolve the present forty-hour week, we dissolve holidays, we dissolve weekends, everyone has a job on a shift schedule of his own choice, for thirty-two hours. We, of course, use up the surplus of people who are unemployed, and put everyone to work in something he truly wants to do.

MA: *Any thoughts on radiation, acid rain, pollution levels?*

BP: Well, like the handgun, homelessness, and unemployment, I'd like to point out that where it is possible that the levels of radiation, acid rain, pollutant levels and so forth, where these levels are measured and known by commercial and specialized agencies, they are in general kept from the public, who in one way or another have paid for revelations about them. Such values in numbers and forms should be, according to us at the Institute, posted daily in the region by the newspapers, just as sunrise, sunsets, and tides are now reported. This is obviously true for anyone within a twenty-mile radius of nuclear power plants. In fact, in such areas, levels of radiation are to be reported by radio twice daily, in translatable terms understood by all. The air pollutions of industrial waste smoke levels must also be reported for the well-being of all. Now, again, I have to admit that the chance of this happening is quite remote, but it is obvious that whatever is reasonable, we're not going to do. I'm merely saying that in the newspaper or over the radio, you should be able to get a reading which indicates the level of radiation at a given hour, the amount of acid rain present, and you should be able to be given the value of the pollutants in the air which you breathe.

MA: *What of this radiation thing? Dirty air and acid rain seem obvious.*

BP: Well, yes, whatever is obvious we're not going to do anything about. Now we come to this atomic power business, we have to bear in mind that atomic power at

HOLD ON TO YOUR HAT

the moment is an uncontrollable monster. We have to recall the history of atomic power, namely we go back to Madame Curie in France, her husband, her son-in-law, her daughter, who ground a substance called pitchblende and reduced it to particle form, and called the result radium. They went on to discover other elements called the radioactive series: that would be radium, uranium, plutonium. How wonderful it would be to just sit around and read about their work in the general physics books.

There was also a woman, a theoretical physicist named Lisa Meitner,[1] who in Germany worked out on paper the formulae that uranium would explode. The uranium, of course, she discovered was divided in two parts and one part of it would explode and produce extraordinary amounts of energy. She brought this to Einstein at Princeton and asked him, would he confirm her results. To his joy he confirmed her results, but to his sorrow he took them to Harry Truman and Truman said, *Well, we will proceed forthwith to make this come about and we will make the bomb and we will drop it on Japan.*

It's very interesting that Einstein and Truman could have gone to the Emperor of Japan and said, *Look, this is what we have, this is what we're going to do: We are going to release this energy from uranium, and not only are we going to release it but it'd be possible for us tomorrow noon to make an island off the coast of Japan disappear.*

Or we could save our people from going to war by just sitting around reading about this. But no, we're going to make it and we're going to draft the physicists and the people of our country who are qualified; we're going to separate uranium which we brought down from Canada, we're going to divide it into its two parts, we're going to put it into a bomb, we'll put it in a plane and order the plane to proceed in a certain range at a certain speed, under orders that the first place that is open below the clouds, we will drop it.

It's very interesting that in this flight of the bomb-carrying plane, the first city to be passed over was a Christian town—Baptist missionaries had come to Japan in the early days and had converted a great portion of the inhabitants of this town to a Baptist way of interpreting the Bible. And it happened that the men in

---

1. Lisa Meitner (1878-1968): Austrian physicist and first woman to receive a share of the Fermi Award. The element meitnerium is named after her.

the plane were also Christians, and the mythology has now developed that this first town that the plane went over was saved because the people below were Christian and the people above the same thing.

At any rate, the plane proceeded, and the next opening was Hiroshima, where from five-hundred meters above the earth, in an instant the place was decimated, as we know. And statistically now, in 1989, persons in Hiroshima are dying because forty-five years ago they were radiated.

So we have the case of Chernobyl, in Russia, where we read about a cloud of radioactive iodine, cesium, that sweeps over the fields, the forests, moors, and the lakes of Sweden, necessitating the slaughter of thousands of reindeer. *Reindeer?* you say. *What's a few thousand reindeer got to do with progress?* Well, in dosages that are high enough to kill them, and have them killed or make them unusable as food, what is radioactivity doing to people? Well, we shouldn't get too inquisitive, because who cares what happens to reindeer or people, especially people, if there's not someone somewhere making some money?

Now, this happens to be one of the forces of terror and horror in our world—the making of money. When the bomb was dropped on Hiroshima, I hope it's obvious that it wasn't necessary to go on to Nagasaki and drop the other one. But the forces that be, the politicians, had taken this out of the hands of physicists, and had gone to war with it, whereas the physicists in Tennessee used to walk the hills, have picnics on weekends, consider some forty-one uses for atomic materials. And of course none of those are being used, not even today.

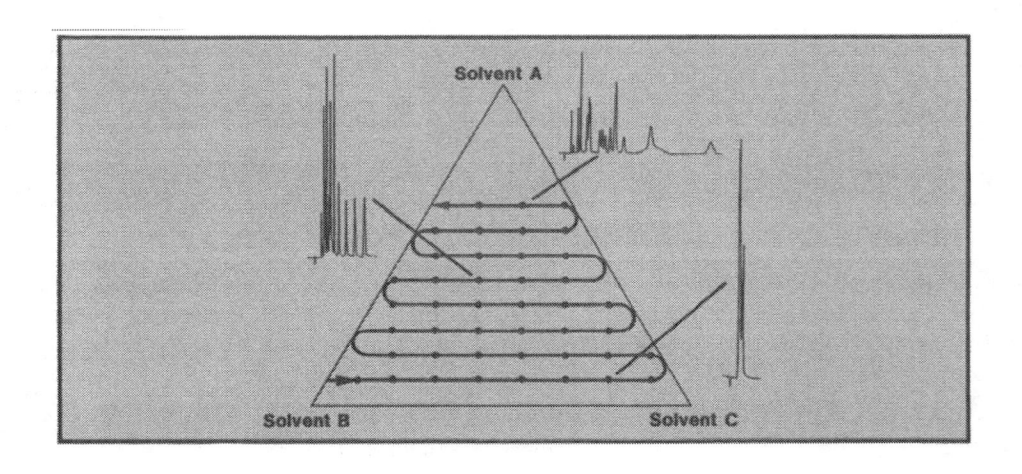

HOLD ON TO YOUR HAT

We carried on a war at incredible, incredible cost in terms of money and deaths on both sides, Japan and our own people, and the deal was simply to, in consideration of a dollar, give everything to General Electric, in consideration of a dollar, to Westinghouse, and to say, *Now we're going to help you.* And Westinghouse and General Electric proceeded forthwith, with great haste, on a moneymaking deal to give the American public power plants, atomic power plants.

In the course of their haste they never bothered to figure out what to do with the waste, they never bothered to figure out the incredible cost in comparison to sun power, tide power. The incredible inefficiency of the whole thing, the breaking down of radiation in people, the irradiation of the workers … just a wholly, wholly uncontrollable monster now in 1989, from this great rush to make money.

So the Institute, of course, in its Theory of Simplistics says, *Well, let's stop all this horror and let's not make any more power plants, let's not create any more waste, we do not know what to do with the waste. Let's look in a fresh way at what we're doing, see if we can't use the tides, if we can't use the wind, if we can't use the sun.* I believe I've heard you say, *What can we do about noise, or whatever can we do about silence?*

Well anyway, with the reindeer and Chernobyl, we've heard in the newspaper of other accidents, and even in Washington State we have what are called *contemporary artifacts*, namely, power plants which were halfway constructed at incredible cost and just abandoned, because there was really, frankly, no more money. And so the truth is the fact that power plants should never have been built in Sweden or anywhere else in the world. If the plants were to be built, and they mistakenly were, then there are other ways, as I have said, of studying the winds, tides, the rain. And I'm very excited about your mentioning of what can we do about noise.

The technique of looking at old things in a new way. For example, on the floor here you have some shoes … You call them sneakers, I don't know if you remember, in the old days it was just a piece of canvas and a rubber sole. You were able to buy a pair for $3. I suppose those sitting there are….

MA: *About $40.*

BP: Well, it's very interesting to me that we still, from the point of view of physics, do not understand the technique of walking. To understand the technique of walking means camera studies of, say, 1,000 people walking by the camera with the camera focused on the foot, on the action of the foot and what it does or does not do in the course of walking; and then devising a container for this. It's very interesting that probably we don't need any containers or shoes at all.

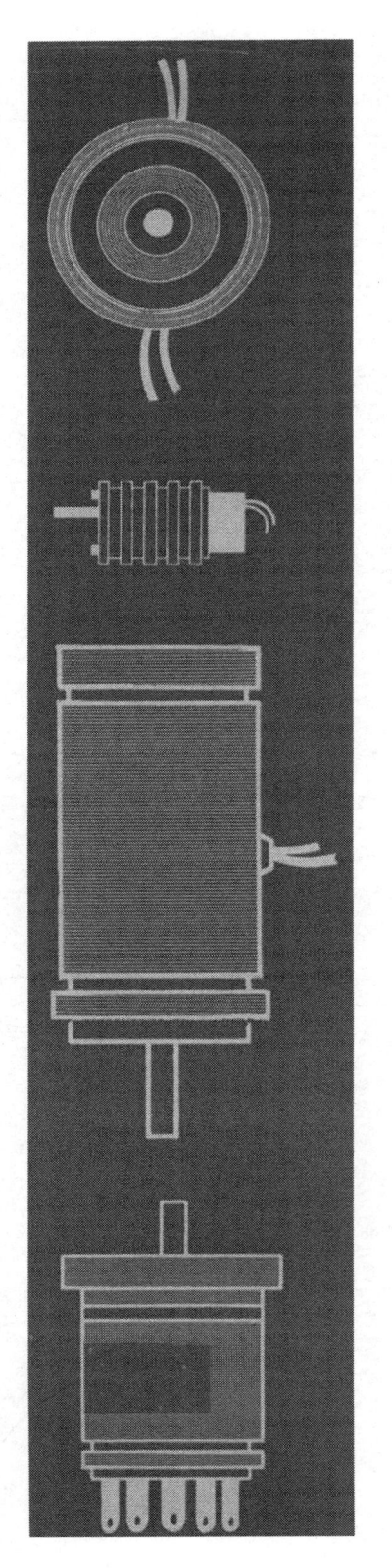

MA: *Just to walk barefoot.*

BP: Right. In that case the sole of the foot would develop its own sole, becoming thicker than the present sole, and it would be able to resist heat and cold, roughness and texture. So what are shoes for? Well, it's quite possible that they are made to make money.

Now, we were talking about your footwear, where it has side bands, it has toe guards, it has linings, it has initials, it has trademarks, it's even available in colors, you can use fluorescent laces, it has tongue flaps, it has grid soles, it has side heels, there's metal cleats, and all of these, as I repeat, have no particular function for the act of walking. It's easier for manufacturers to forget about the act of walking and decorate until the shoes cost you $40. As opposed to the old days when you bought a simple canvas and rubber sole, or the days when you just went barefoot.

So do you have another instance you'd like to talk about?

HOLD ON TO YOUR HAT

MA: *How about books?*

BP: Well, this is very interesting. You and I have produced a book video. What is a book video? Well, it's a recognition that people spend interminable hours sitting before screens. It's said some people spend up to six hours before a television tube, which as I've said should've been used to educate rather than to entertain.

And it's an admission that people, probably in our time, are just so involved with the stresses and strains of what we call the rat race, that they don't read anymore. But they will sit down before a screen. So we in effect have provided them with a system where they don't have to think, all they have to do is sit, pages are turned for them before their eyes, and the imagery conjures up in their minds from their past experience all the facts which they had earlier obtained by the act of reading.

We are in effect saying, people no longer read, so what are we going to do about it? Well, we're going to give them a book-video, which you and your concern here in Madison have already done. In other words we have produced a book for the future. This book for the future is something which is projected by electronic means onto a screen. You will no longer turn the pages, no longer read in the sense that you used to.

MA: *And what about words? You think words are about to disappear as well?*

BP: Words are also on the way out, and it's interesting that in the course of reading one has overlooked the true meaning of a word. Many of us have heard words but few of us have ever heard their meaning, their sense and their music. Many words are now so bruised in our culture that only their music remains.

So the Institute now requests that we try now suddenly to listen to the quality, to the spirit, to try to understand the feeling that is transmitted. In fact if you think about it, you find that for two people using words, it's amazing that they can understand one another at all. The experience and training and background of each is such that a single word uttered has several meanings between the two people. So that you can be using one word in its sense to you, and I can be using the same word in its sense to me, and we do not reach one another. In fact that's one of the common horrors of our culture, that we are unable to reach anybody, we don't touch anyone. We certainly can't get through to them with words, so we're separated and encased in an armor, as Wilhelm Reich suggests. We're in cylinders about six feet high, three feet in diameter, and these cylinders, we're in the center looking out through swinging doors, swinging windows, and peering at one another rather than touching one another, getting through with words or feeling or spirit or anything. We are, as the Institute says, making more complicated what is complicated, we're not simplifying and getting around to the act of feeling and understanding through feeling.

So yes, words are on the way out. Have been for some time.

MA: *Is there an international way to communicate feeling?*

BP: Several techniques have been used. One is quite common in Indian culture: a man will sit under a tree for forty years and contemplate, try to project his body beyond the five senses into a sixth sense. Another method, one I've advocated in SciPoe,[2] is to reduce the whole thing to dots and dashes, and flashes—to use light, and radiated fluorescent materials and methods.

So we're in a state of change and we're again looking at old things in a new and fresh way.

MA: *Is there a problem that if you begin to simplify things too much that you will sort of lose the essence of the world as we know it?*

BP: Well as I say, in our culture we tend not to simplify, and your question is, can we simplify to too great a degree, and lose the core of the meaning? The answer is no, because things have what in physics is called a *plasma*, and a plasma is an energy, and an energy can not be destroyed. It can be felt, cannot be seen, cannot be bruised, cannot be cut. It is fluctuating, it is in a sense a *vibration*, and all things can be reduced to a vibration.

---

2. See Porter's *Physics for Tomorrow*, reprinted on page 185.

HOLD ON TO YOUR HAT

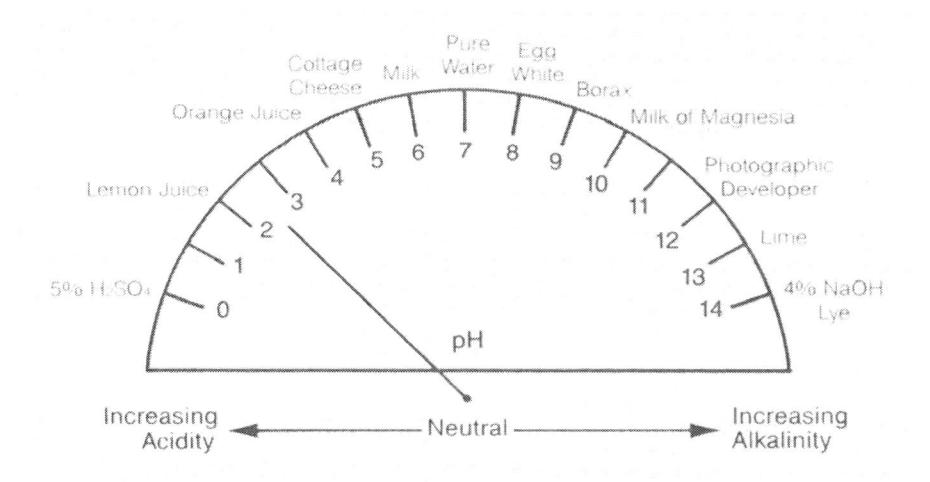

MA: *Which is truly intergalactic.*

BP: Well, this brings on the universe, when you use that word. And what is the universe? Well, it also reduces to plasmas and to energies and vibrations.

MA: *When I was a kid, when I was in chemistry class and I learned the periodic table, I put together this theory. Our solar system has a sun and nine planets, and the lithium molecule also has nine planets around it. My theory was sort of a micro/macro theory where our solar system, the sun and the planets as we know it, is actually a lithium particle for a larger world. And in that way I was able to go infinitely in each direction, where then that would be a lithium particle for the next thing; and you can also break it down in the opposite direction where the lithium particle that we have in our planet would actually be another solar system, with all its various complexities.*

BP: Well, I would like to congratulate you on this because it is quite fundamental and quite feasible. What we're in effect saying is that nature provides for itself in all respects, and duplicates itself in many variations and many forms. The base is maintained throughout. And in your case the lithium is a common ingredient which will appear throughout all systems simultaneously.

Interviews with Bern Porter

MA: *Do you think nature is entropic?*

BP: Well, nature, we're still in the process of studying it. It's very interesting that we're separated, just as people are separated by armor and in their cylinders looking at one another; nature somehow refuses to allow us as observers to know everything about it. It's like we are separated from the True Fact by a membrane. And we push the membrane with the right hand, then the left hand side comes out and hits us in the face. And this membrane is of such extent that we cannot go around the ends, we cannot go down under it, can't go over it. All we can do is push it, push it forward. But in the membrane there are swinging windows; swinging doors which open in, and then open out. And our problem is, can we look through those windows? And the answer is, we cannot. We cannot completely get through it. We cannot even see momentarily, we can merely be in the presence of the windows; and know that on the other side is the ultimate answer, which will never be revealed to us. We merely make stabs at trying to understand phenomena, and understand the making and the design of this tremendous, wonderful thing called nature.

MA: *How do what are known as aliens or UFOs fit into nature?*

BP: Well, it happens that all of the elements of nature are plasma, are energies. It happens that there are some energies yet to be identified, and among them are these, what we like to call Alien Bodies.

It's interesting that if you were to magnify a point on any solid by say 500,000 times, you would find in the photograph that the so-called solid part (which you thought was solid) is really made up of mountains and valleys with spaces in the valleys and between the mountains, and you would find incredible spaces; and in those spaces of course are energies, and these could be called (not only *could* they be called, but they are) alien because they are unseen by us. But they can be seen by extraordinary magnification, they can be felt and interpreted.

HOLD ON TO YOUR HAT

MA: *And these are entities with an intelligence?*

BP: These are energies within nature, which implies intelligence, yes.

So we have then a complex mass of energies which are operating simultaneously. And it's very important to understand the word *simultaneous*. It means things are going on at the same time, a great number of things. As observers we are unable to comprehend or understand all of them as they take place.

But it's interesting and very important to realize that these energies are proceeding through space. They themselves have three dimensions but they are effected by a fourth dimension, called time. Time is flowing through them, they are flowing both against time and with time, time is ever present. And this simultaneous action is going on in such a way that there's actually no past, there's no future, there's only now which we call the present, and we are proceeding then from one second to the next second, through this time frame, having as we proceed with time and against time three dimensions: height, breadth and thickness.

And as we proceed from one second to the next then the only thing to get us through are our five senses. And the Institute keeps pressing in all of its work that we must take care of our five senses. It's a rare individual who would use five at one time or even three at one time. And the horror of it, of course, is that our hearing is going, our seeing is going, taste is gone pretty much. And evolution-wise it's possible people will be born, say two-hundred years from now, without the faculty of taste, no longer tasting anything. And we're losing the sense of smell.

MA: *But are those being replaced by other senses?*

BP: Well, man has it in his power, if he so wishes, to combine his five senses, to get onto a sixth and seventh. But it's quite possible that in this system of not simplifying, making complex, or simply just making horrible messes, as we see on all sides, we will not do this.

MA: *So we might be reduced to no senses, is that what you're saying?*

BP: We very possibly, at the rate we're going, will be reduced to no senses whatsoever.

MA: *And what would that leave?*

BP: Well, that would leave us as a speck of mud I suppose. Without intelligence, without bearing, without understanding, too horrible to contemplate—that evolution-wise, biology-wise, we are somehow born without any senses at all.

But it is true that our hearing is going, it's braced by hearing aids; our seeing is braced by glasses. We don't taste any more. It's interesting if someone asks the question, why is our nose so close to our mouth? And the answer is that the mouth is supposed to be tasting and the nose is supposed to be smelling, and doing the two acts simultaneously, but they don't. So maybe we should redesign the face. Maybe we'll...

MA: *Put the nose further apart from the mouth?*

BP: Yes.

MA: *Maybe on the back of our head or something like that?*

BP: Or have no nose at all. So Nature has provided for everything but we have bruised, and not only bruised but we have re-modified, and we've done horrible, horrible things not only to everything but just everything. Our environment, our bodies, our minds, our intelligence, everything is slowly disintegrating. It's a horrible thing to contemplate. So now what is the Institute saying? It is saying we now must save what we have, we must proceed immediately to clear up all the problems, at incredible cost.

There are spots in the United States where we've been throwing out industrial wastes for years, and now suddenly we find that industrial waste is getting down into the water system, contaminating the water; it's been contaminating the air for years. Now we have to clean it up, at incredible cost, when we should never have made it in the first place. That's the story of atomic power plants, that's the story of industrial waste. Incredible number of flaws in our culture. And of course on top of that we have man and his greed. And we have this thing with the savings and loan situation where we have incredible costs to recover, incredible costs to restore. So man in general is not a very pleasant creature to contemplate.

MA: *What does the Institute think about money?*

BP: Well, we find it is an uncontrollable monster and we would much prefer a barter system.

For example, I was very impressed to see on the library board in Belfast someone who would like to exchange vegetables grown in their garden for books. This is an interesting example. I have books which I have published, and I certainly have some need for vegetables. So I simply go to this person and he gives me a peck of potatoes and I give him three books of his choice.

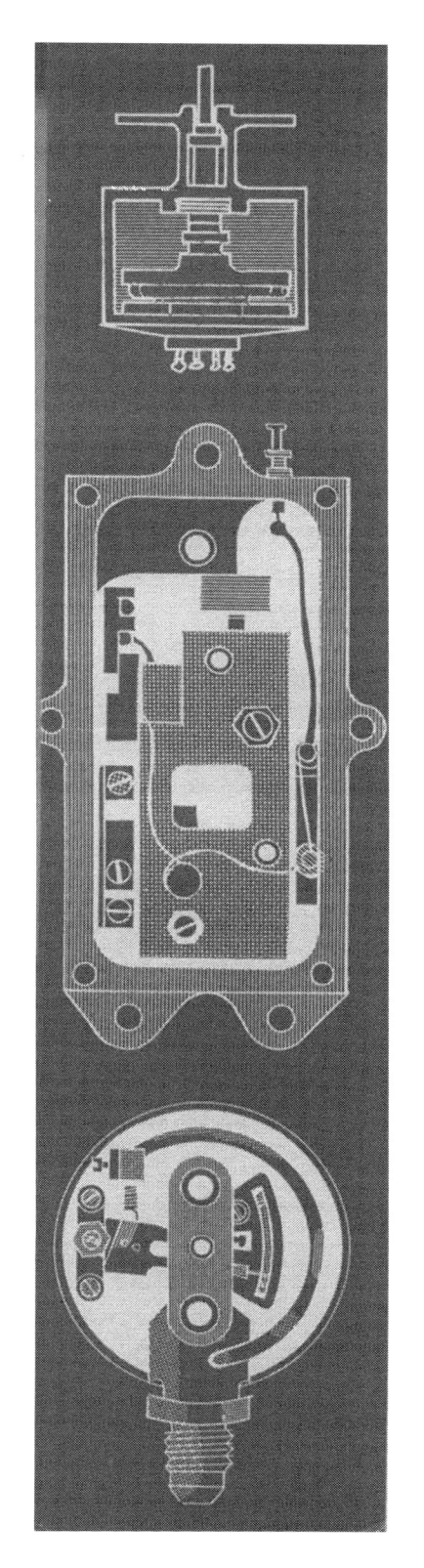

In one of the most satisfying transactions I ever made, I went to a farm where they grew sheep, and they sheared the sheep every year and made rugs of sheepskin. And I thought how wonderful it would be if I could have a sheepskin to put on my bed as a mattress, and how could I get it? *Well*, the man said, *You have some posters with you that you have created, in your suitcase. You give me three of your posters, which I will hang on the wall and enjoy; and I will give you one sheepskin which you will have and enjoy.* And this was a very satisfying transaction. He became very happy with his posters, I became very happy with my sheepskin, and we did not pass any money, did not use any money or checks or forms.

MA: *But isn't it because of money that you're able to establish a value for each of those things? How do you know what to exchange for what?*

BP: This is a matter between the individuals. In the case which I just gave, the man said, *Well, I want your three posters, it does not matter what they cost, it does not matter what the sheepskin costs.*

Now secondary to the barter system is so-called *barn raising*. A man in a rural area could say, *Well, Tuesday I wish to build a barn.* Word gets around that so-and-so is going to build a barn on Tuesday, and twenty or thirty men and their wives show up, the wives prepare food and the men proceed to build a barn. There's no money exchanged, it's a party, it's a combination of party and work. When the barn is built, why, the owner of the new barn, he goes down the road the following week and he gives his labor to a second neighbor to build a barn.

So here's a system where we're sharing labor, and we're not paying someone $12 an hour to build a barn. So there are systems which have been demonstrated through time, in rural areas and native cultures, primitive peoples we call them, who are involved in survival and find that by barter they can exchange things, and by sharing work they can survive.

Coupled with that then is the apprentice system. I have a brother in Maine who wanted to be a pharmacist. So what did he do, well he went to a local drugstore, told the pharmacist he'd like to learn pharmacy, and he started off sweeping floors. The pharmacist gave him books to study at night, and taught him there on the job how to be a pharmacist. And my brother then went to the state, took the examination and the first time through he flunked it, and I understand he flunked it the second time, and he went back a third time, and became a licensed pharmacist.

Well, this was a common practice, this apprenticeship, and like the barter system that I speak of, like the exchange of labor that I speak of, the apprentice system is gone. In those days the blacksmith showed his son how to be a blacksmith, the pharmacist showed my brother how to be a pharmacist. Those things have gone, we feel now if you want to be a pharmacist you go to pharmacy school. If you want to be a carpenter, you study carpentry, go to a technical school and learn the acts of planning and sawing and cutting and hammering and joining and all the rest.

These are acts which can be passed from one person to another, and you say *where does value come in, where does money come in*, and I'm just pointing out that it's possible to survive without money.

A very interesting further example of this is the native people in Maine lived there some 10,000 years; they didn't have money. They lived off the soil, and they lived from the animals and moved where there were fish and seafood in summer and there were deer and bears in winter.

So yes, there are cultures that live off the land and who do not have money as we know it. Money has turned out to be a thing which you want to invest and to grow interest and pension plans, and as I've said a great many are still unemployed, and you have to throw out the whole system.

Everything that you know, the Institute advocates that you look at it in a fresh and new way, and generally just throw it out and start over again.

MA: *What if as my chosen vocation I wanted to be an avant-garde poet, and it was something which on a common everyday basis does not have value to very many people. How would that work?*

BP: Well the first impulse is to say why don't you go to University X and take the course numbered 23B, which is Creative Writing, and why don't you get a book on how to write a poem, and why don't you sit down and write a poem. But these are merely substitutes for the real thing. To be a poet requires, again, this relation to the five senses, using them completely, it requires seeing old things in a new way, it requires seeing and feeling and getting the spirit out and getting beyond the work. Coloring the spirit of the spirit. Or as Gertrude Stein would say, a rose is a rose is a rose is a rose is a rose, and you do this 'til you feel it, and having felt it, you express it, and then how do you express? Well again I would prefer that you do this with lights and dots and dashes, but if you want to you can put it down in a word, and you have shown in some of some of your recent work that you've been able to *devise* words.

  And this is what [Abraham Lincoln] Gillespie[3] did, he combined old forms and created new forms. And it's interesting that he didn't go to school to do this, that he didn't read any books to do this, and he didn't study. He merely felt it inwardly and he outwardly recorded it.

MA: *I think what I'm asking is how do I make a living?*

BP: Well I just said you give your friend down the street six poems and he gives you a loaf of bread...

MA: *But what if he doesn't want the six poems, what if he has no use for those six poems?*

BP: Well, in this case we have to change his point of view and your point of view, and demonstrate that, first, bread comes from flour, flour was originally grains, grains have been ground, reduced to flour and have gone through the baking process. And you have to prove that your poem also went through an equal process and so has just as much value as the bread.

  Well, there are people who will laugh at this, including you, and you will say this is impossible, but I think that this is what we're going to have to come to, whether we like it or not. And again, this is a case of looking at old things in a new and fresh way. I still think it's possible to exchange six poems for a loaf of bread.

  And you can say well, what if he doesn't want the poems? Well, it's very interesting that he could take his six poems that he got from you, six poems that he got from someone else, and he could combine them if he wanted to and make a book to sell, and use it for barter. I have great faith in the barter system, the apprentice system, and the use of everything around us for our survival. This is what the Institute is advocating. And this incident is what's called advanced thinking. Advanced thinking: there's nothing advanced about it. It's simplifying what is

---

3. Abraham Lincoln Gillespie (1895-1950): American poet. See The Syntactic Revolution, ed. Richard Milazzo (New York: Out of London Press, 1980)

complicated, and cleaning up what are messes. And we have an incredible number of them. And incredible misunderstandings. We have the wrong system! The system shouldn't be based on money, it should be based on barter.

MA: *Or feeling.*

BP: So what is the next question?

MA: *Well, I was wondering if maybe, to do a little sidetrack here, you could talk about somebody else who's a member of the Institute and some of the things they've contributed.*

BP: Well, we have one man who's very much concerned about gravity. He would like to start off and say, *What is gravity? How can I attack gravity, how can I get rid of it?* And in the course of asking these basic questions—what is it, what's it do—he has found, contrary to Newton and Einstein and some others, that it's not a single force. Newton dropped an apple—held an apple out and let go of it—it fell. He felt that he'd figured it out, and we've sort of accepted it through the years that there's a single line of force pulling the apple down to the ground.

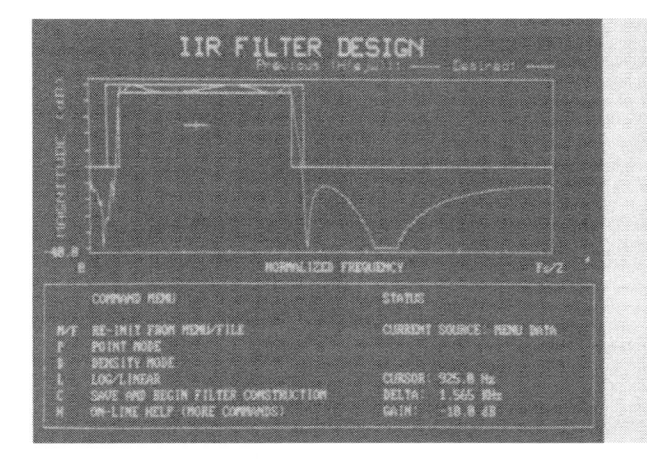

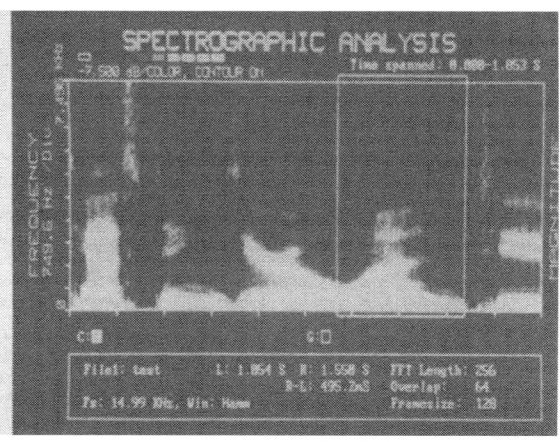

    Well, naturally our friend at the Institute has discovered that there was not a single force but a group of forces. And he also discovered that there is such a thing he can create, which he called anti-gravity, where he can stop the pull of gravity. That is to say that he can inject a plane between the object falling and the ground to which it falls, and produce an effect called anti-gravity. So he studied this, with the help of the rest of us, and might make a contribution. There is anti-gravity, and we can defeat gravity.

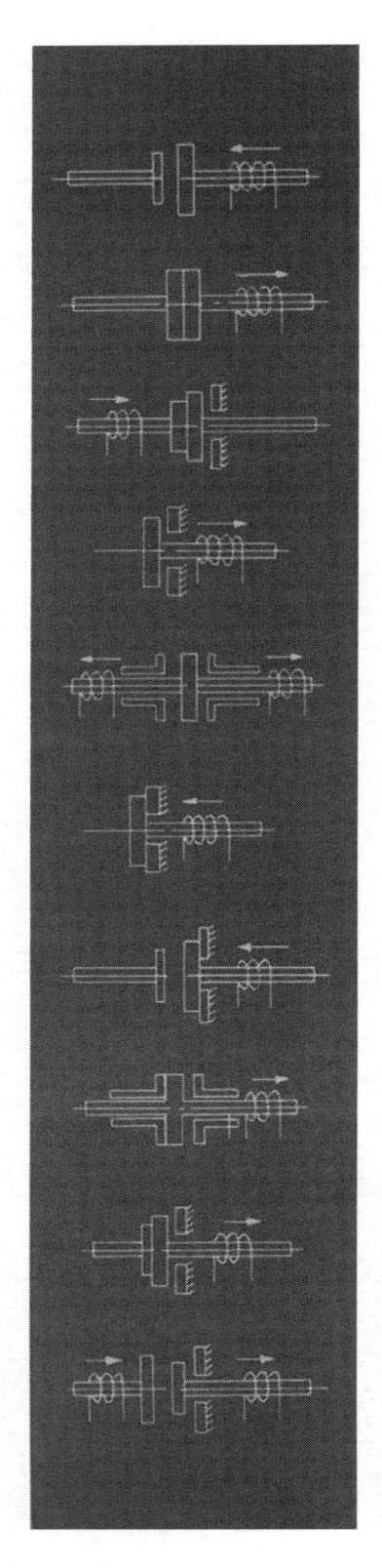

MA: *To what purpose could anti gravity be put?*

BP: Well, obviously things will no longer fall, we can stop things from falling, or stop things from rising. It's interesting that the missiles that go to outer space have to defeat gravity. Having defeated the Earth's gravity, then it's very interesting that their further success depends on the pull of gravity of surrounding planets. So gravity is around but we have to understand it, and we have to study it, as some members at the Institute are doing. You ask me, *Well, what are we going to do when we overcome it?* I think that is part of our continuing study. We have great hopes that anti-gravity will have uses. And it would be very sad if the politicians got a hold of this, and bear it off to war; or the moneymakers find gadgetry that they could sell.

MA: *Do your members ever cooperatively arrive at an idea, maybe one person thinks of part of it and gives it to another person, and they think of another part? Almost like an Exquisite Corpse in a way, of just taking an idea and passing it from person to person.*

BP: Well, that is the function of the headquarters in Belfast where I directly receive these ideas from the field, so to speak, from the Institute's thirty-one members. And I redistribute them in Xerox form and ask all those who would like to continue with this particular idea, please do so and send in the results of your work. So we have projects which are going on for many years.

For example, we're very much concerned about orgone energy, which is the energy of living. Anything that's living is giving off energy and this energy can be measured. It's possible, for example, to put a couple of straps around the trunk of a maple tree, say, a foot apart, and bring wires from the metal straps into a galvanometer. The galvanometer needle will deflect, showing that electricity is being generated as the tree grows.

So we would like to get a hold of this growing energy, the energy of growing, which Wilhelm Reich called orgone, and develop it for physical use. He, for example, was able to direct it with a tube, one end of which was connected to flowing water—a stream, say— and the other end open. He was able to direct this to the clouds above and to deconcentrate the clouds in such a way as to give up their moisture content and make it rain. So this is an example of how energy can be directed to a purposeful use. And you can say, *Well, supposing we don't want rain today but we want it next week, or we don't want too much rain, we just want a medium amount; how are we going to control it?* And again, we're working on an answer.

So yes, we're concerned now with anti gravity, we're concerned with the meaning of orgone energy. How can it be used? And I would like to compliment you on your thought that noise can be reconverted, and hope you will continue thinking about it. With your permission I will probably steal the idea from you and pass it on to our members and see what we can do.

MA: *Now, can ideas be owned? How can you steal an idea?*

BP: Well, this is very interesting to me, the copyright system and whatever. There's nothing new under the sun. While you're sitting here, you and I are talking about anti-gravity and useful noise, there's someone somewhere that has already done this and thought about it or is now thinking about it or studying it; it's not possible to steal anything from anybody. It's universal, it's in the air, it's available to anyone and everyone, and that's why we don't patent anything or copyright anything.

MA: *At the Institute.*

BP: At the Institute. I said I'd *steal* this thing from you but while you think your idea is original with you, I would point out that somewhere someone has already thought about this and they could conceivably be working on it and developing it to a very high degree.

MA: *How does one become a member of the Institute?*

BP: Well, the requirements are quite simple: on one side of a sheet of paper, if you cannot express what it is on your mind, there's something radically wrong. It shows you haven't completely thought it out. And having submitted it, it in some way must involve physics, and you yourself cannot be a member of the faculty of a university, you cannot be in the research laboratory of a private company, and you cannot be working for any government. You have to be a freelancer.

And the question comes up, *How do you guys support yourself, get through the business?* Well, we're trying the barter system and sometimes it works, sometimes

HOLD ON TO YOUR HAT

it doesn't. Quite frankly we're all freelancers, and we derive our income from—one member of the family may get up every morning and go to work, support the one who's staying home to do the thinking. What is thinking? Well, it's a process of resolving the complicated to the simple. It's a process of seeing old things in a new way. And it's also important to consider things as functioning simultaneously in time.

So how do you get in? Well, you submit your scheme on one piece of paper, and you come to Belfast, Maine between May 1 and November 1, and you bring with you a tent and sleeping bag, and you meet with others there who are also in tents or in sleeping bags, on platforms. You discuss your idea or series of ideas. It can likely be found to be related to physics in any way—it's interesting that physics is germane and basic to a staggering number of disciplines. To be an optometrist, a veterinarian, an engineer or a biologist—or a whatever—at some point or other you involve yourself with physics.

MA: *Has anyone ever applied to the Institute, then been rejected?*

BP: No one has ever been rejected. We merely say, *Well, you have some ideas, please come and stay with us, please keep in touch with us and please report to us how you are continuing with your idea.*

It's very interesting that people who have ideas are usually beaten. I guess that's the word I would use. Their mother or their father doesn't understand, or their brother doesn't understand, or their relatives or people around don't understand what this person is about and it's very easy for society to say, *Well, here we have a crackpot on our hands, another guy who's half-insane.* And it's interesting that the borderline between sanity and insanity is very thin. Indeed they may be one and the same thing. It's unfortunate that society has to label and classify, put all of us into little pigeonholes, and not only classify us, pigeonhole us. They would like us to stay there, to stay in the mold and be accepted. And then having been accepted we are expected to join something, to be a member of the Republican Party, of the Democrats or the Baptists or the Jehovah's Witnesses or the Oddfellows. We're expected to be joiners; it's expected that we fit in. And if we don't fit in and we refuse to be classified, then we're called crackpots, sub-geniuses, or we're given all kinds of labels. Which is very unfortunate, very unfortunate.

MA: *Aren't there any crackpots from history that are sort of influences on the Institute? People from past ages?*

BP: Well, yes. In our time, of course, there's Buckminster Fuller.

MA: *What about in past times?*

BP: In past times there were a great number of them.

MA: *Are they people that have been remembered?*

BP: They are still remembered. Leonardo da Vinci created innumerable devices, and his devices and ideas were forerunners of the airplane and of oil drilling equipment, and he conceived of a staggering number of things, but we still say, *Well, he was just a backdoor inventor, just a side guy.* We rarely give credit to anyone for anything. Except maybe persons who've made a great deal of money.

So yes, in history there are bonds with people who we've thought to be—well, we've been concerned the last few days with [Abraham Lincoln] Gillespie. He took off where Joyce and others left off, and died unknown, for all practical purposes. I personally have tried quite desperately to see what I can do to revive him, at least mention his name around.

[end of tape—conversation moves to Wilhelm Reich]

... a man who in Norway and Denmark tangled with Freud and others in the field of what was called psychiatry. They ostracized him, he came to New Jersey and finally came to Maine, where he built his orgone accumulators, his rainmaking machinery and whatever. And what happened, well . . .

HOLD ON TO YOUR HAT

MA: *He died in prison, right?*

BP: The federal agents came by and claimed that his accumulators were not curing people. To my knowledge he never claimed that it cured anything. He merely made them and let the users decide for themselves what it would do or not do. At any rate the authorities came in and took his equipment out in the yard and chopped it up with axes, set matches to it to burn it, then they brought out his books and burned them to ashes, took his books, and took him off to prison, where he died.

He left behind an incredible array of original documents, which in his will he testified should not be allowed to be interpreted directly until after fifty years had elapsed. His argument was that, *Well, in fifty years maybe people and society will be able to catch up with what I was thinking about, and will be able to confirm what I've said.*

Meanwhile in spite of the will there are people all over the world, including members of the Institute, who are pursuing his ideas. In fact, at the Institute we have had for years an orgone accumulator—a horizontal one measuring about twelve by eighteen feet. It sits back midway between our property, and is free from the electrical power lines out on the streets, and is sitting over a free-flowing water field, which sadly has been since disturbed and destroyed. But it did function and as far as I know is the only horizontal accumulator in existence.

And a person would go there and I would never claim anything, other than the fact that when you step on it and go on to the center, this twelve-by-eighteen-foot flat surface, sit on it or lie on it, you would feel something, and what you felt was up to you. What it did to you or didn't do to you, I didn't claim anything, and neither did Reich. But the authorities said, *This is quackery. You cannot get away with this. We will burn your books, we'll put you in jail.*

But this is a very fine example of society closing in on a man of ideas. And we have yet to prove that he was right or wrong. As a matter of fact we're still looking at Einstein and all of his notes. Einstein developed his ideas between the ages of sixteen and about twenty-three, and spent the rest of his time from twenty-three until his death just embroidering his original ideas. Of course he died sad and disillusioned and angry, frustrated, and horrified to think that his ideas were turned to war. He said that if he wanted to do it over again, he'd be a plumber, because plumbers do something constructive.

So yes, we have examples in our contemporary history. You want me to name the older people? Well there was Newton, there was Galileo, there was Leonardo.

MA: *What about the alchemists?*

BP: And the alchemists, they were all crackpots. But it's interesting in 1989 that some of their ideas, quite an amazing number of their ideas are still valid.

Then we have what are called *old folk tales*, remedies, cures. And of course I've already spoken about the native peoples in Maine for 10,000 years and their incredible contributions that just came from the soil. So how do you make a synthesis of this? Well that's what the Institute does. There are two approaches: You can take up a subject, and go to a library and research, find out everything that has ever been done about this subject in the past.

Or, you can take the path which we recommend. You forget about all the past, and take off into new fields or what we call breakthroughs. Just apply your five senses to this subject, forgetting everything that's ever been done before, and look at this subject from a new point of view. Namely using or applying your five senses to this, this force in time.

It's very interesting that sleep and dreams are very much a part of this. In the subconscious at night, turning over these ideas, these views and these opinions, and putting forth, in the morning hours, incredible synthesis and analysis.

MA: *The Aztecs believed that their dream life was their real life, and their awake life was their dream life. Have you ever considered that sort of thing?*

BP: Well, that's what I just got through saying, was that sleep ... it's unfortunate that so much of our life is spent awake. It's interesting that the greater part of our work, conscious work, is done during sleep.

Take the case of a man like Salvador Dali. His trick was to get in bed, pull a paper bag over his head, and put some notes on his knee. In the darkness of the paper bag he would sketch out folded clocks, disintegrating trees, exotic landscapes and the whole field of surrealism. These came as mental visions, and at night he would wake himself up and make sketches from dreams. Reinterpreted dreams. Practically every painting he ever did was originally a dream. During the day, to induce the aura of darkness, he would use a paper bag over his head.

He's a living example of how the subconscious and the mind at sleep produces incredible imagery and ideas. So the Aztec view is very valid. That's what the Institute is saying: please do not overlook all of this methodology and all of this culture. Two things: Either study it completely, or take off on your own. And when you take off on your own, become a pioneer and make a breakthrough.

It's very interesting that what you have produced has already been thought of and already been distilled by other cultures and other times. In effect, we have this statement that there's really nothing new under the sun. Everything has already been done by someone somewhere.

MA: *Isn't it a contradiction, that one of your principles of the Institute is to try to do something new, yet you believe that there can be nothing new?*

BP: Well, in a sense you're right, but we always have the next layer up, which is called the new-new.

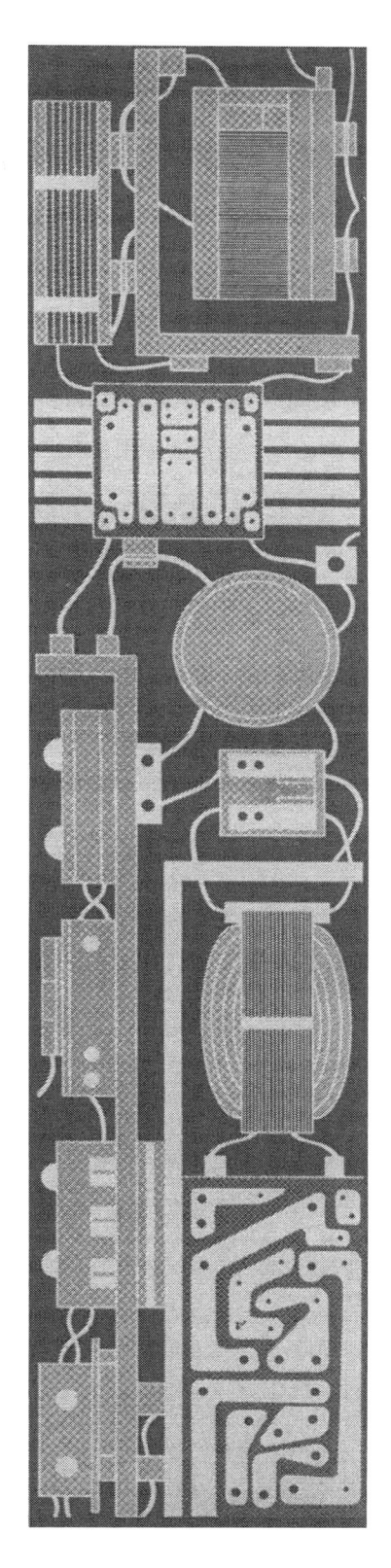

MA: *The new-new?*

BP: Yes. The next layer. And this comes, as I've been saying, through sleep. So we have what is prevailing now, and it's pretty obvious that most of what prevails now is pretty horrifying. And we call it new. Your sneakers on the floor are new, but they're also a total admission that no one yet knows how to walk. So, a new-new for walking or for shoes would be possible. And it may be they can't improve on nature, so you go around bare.

So there are contradictions, yes, but these contradictions neutralize one another. And whatever we think of as new or different somehow has another layer of interpretation.

MA: *Does the Institute have any information on the Tower of Babel?*

BP: Well, the Tower of Babel is a very interesting example. Biblically and historically it is said that the Lord, in setting up races and people, didn't want them to talk to one another. Or communicate with one another. And of course we at the Institute think

HOLD ON TO YOUR HAT

this was a horrible decision. Instead of saying, *Well, you people must have innumerable languages*, it was *You should have one language.* In the same sense, we at the Institute say, *The countries of the world, through the United Nations, should have the same monetary system. There should be no visas, and there should be no boundaries, and all people are free to travel, but have the same language and the same money.* But the Tower of Babel was set up in contradiction to this obvious and reasonable situation that I speak of. It was declared that all people should be divided—first by color and then by speech, and this is a very unfortunate thing which we probably will never be able to overcome.

MA: *In a way, it's responsible for a lot of the problems that the Institute is trying to solve.*

BP: That is correct, that is correct. There are many horrors. One I've spoken of is atomic energy. Another is the disappearance of the apprentice system, the absence of the barter system, the continued pollution of the air and of the water. All of these things are pretty well out of hand by now, with the result that the Earth's lower levels are warming up, and seas are rising, land is being submerged, volcanoes and earthquakes are occurring, and we really upset the whole balance, the whole harmony of just about everything. And it's true, the Institute devotes itself to correcting this. And while it's maybe an impossible job, we don't think so. It will take time, but we think there's still hope, even though it's pretty big.

MA: *How will the Institute carry on into the next century?*

BP: Well, the average age of our thirty-one members … I'm already seventy-nine, and one of our members has died. You might be interested to know that he got into the Institute because he was a photographer and he made his own camera. He ground his own lens, made his own chemicals, made his own solutions, and he developed a system of photography with his own camera … made his own chemicals … where you would encompass a scene and press a button just once, and that's it, with extraordinary results.

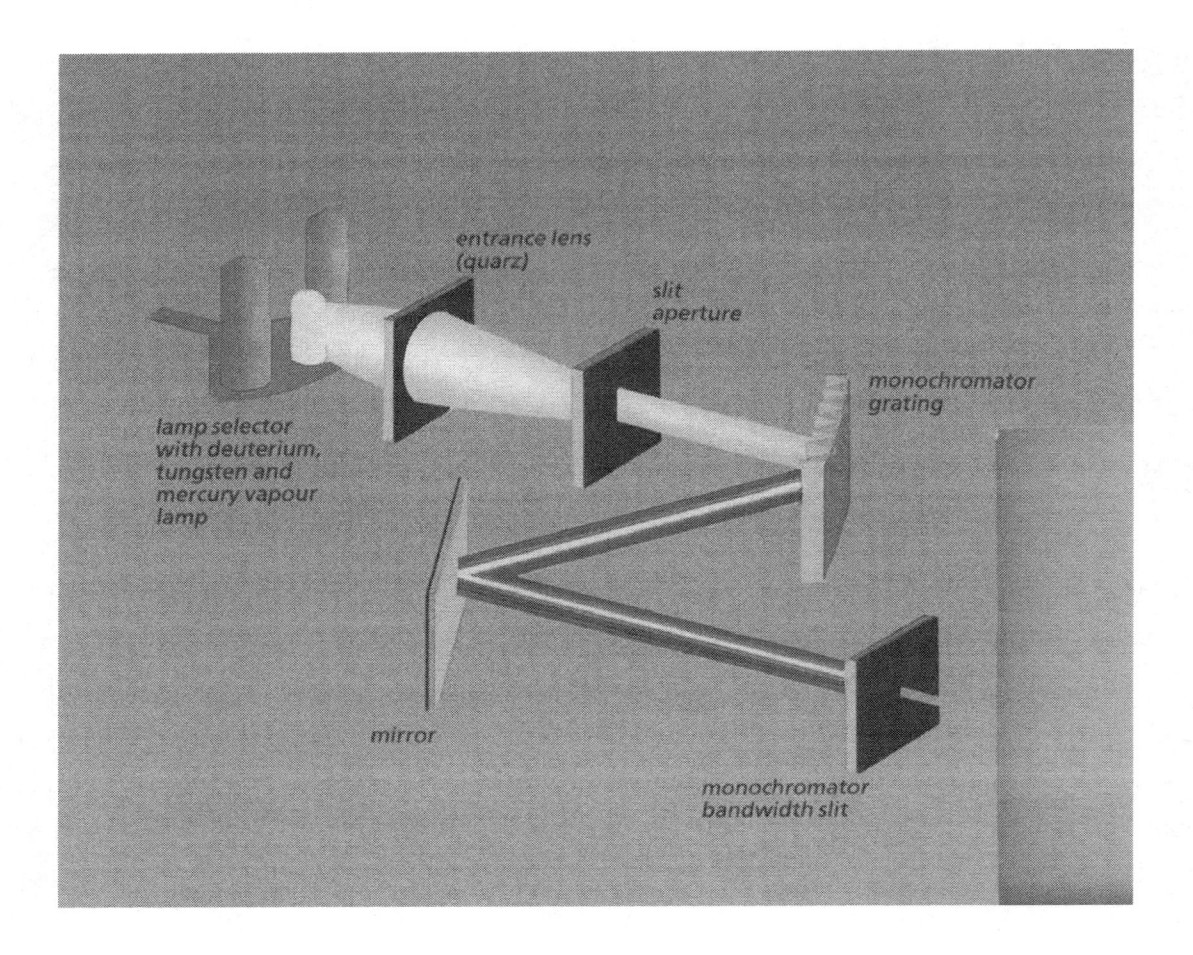

[Diagram of a monochromator with labels: lamp selector with deuterium, tungsten and mercury vapour lamp; entrance lens (quarz); slit aperture; monochromator grating; mirror; monochromator bandwidth slit.]

MA: *Who is that?*

BP: His name was Harry Bowden.[4] He's still on the roll as a scholar, and all of our scholars have fields of endeavor; he was called a scholar in photography. Now that he's deceased, he's never been replaced, for the simple reason we have never been able to find a photographer who was his equal. We have incredible photographic equipment, made in Japan and Germany. But we don't have anyone showing up on the scene who grinds his own lens, makes his own whatever, and produces incredible photographs. So we have Harry Bowden (deceased), scholar in photography. And there's no other who can touch him. Harry Bowden.

Well the average age ... I believe, as far as I know, I'm the oldest, and as I've already said, it's interesting too, with all of the things we have to contend with, here <u>am I, seventy</u>-nine, my bowel system doesn't act like it used to, my urinary system

4. Harry Bowden (1907-1965): American painter and photographer.

doesn't, I'm losing my teeth, it's possible I still will have to use glasses, I still can hear, I still can taste, but old age is coming upon me, and it's not a very pleasant prospect.

But the average age of our members, going on to the next twenty, thirty, fifty years, is such that our ideas will prevail. And of course we always have the possibility of someone showing up and applying, and qualifying, but so far that's not been possible. But that someday someone might. So I have no fear about the future.

MA: *I wonder what the best way to share the passage of ideas into the future is.*

BP: Well the old system was from father to son. In our case we pass them amongst one another, and then we give interviews to the press, and do personal appearances for groups. In Albany I was able to talk with people about plasma. One of the situations I called to their attention was that here were forty of us in a room, and that our energies were concentrated in that room. Each one of us brought in his plasma, from his knowledge, his background, his experience. And we were all crossing our paths, individual orbits going through time and space, crossing one another and creating in that room a tremendous field that I could feel, and I'm sure some members of the audience, after I called their attention to it, could also feel it.
So energy transmits, and plasma transmits the ideas.

MA: *Plasma transmits the ideas.*

BP: Yes. You see, the plasma was always here, and will always be here. At the time you were conceived in the womb, a name was assigned to you, and you proceeded on an arc through time and space. And at your death, what you call death, the skeleton and the container drop away but the plasma continues. Plasma never dies, and is ever present. And when a group of people assemble, then, well ... you and I here in this room. All your experience, knowledge and background fuse with all my knowledge and background. And it's possible, since you are sitting here vibrating, for me to tune my system to your frequency of vibration, and for me to speak to you without using words, or uttering sounds.

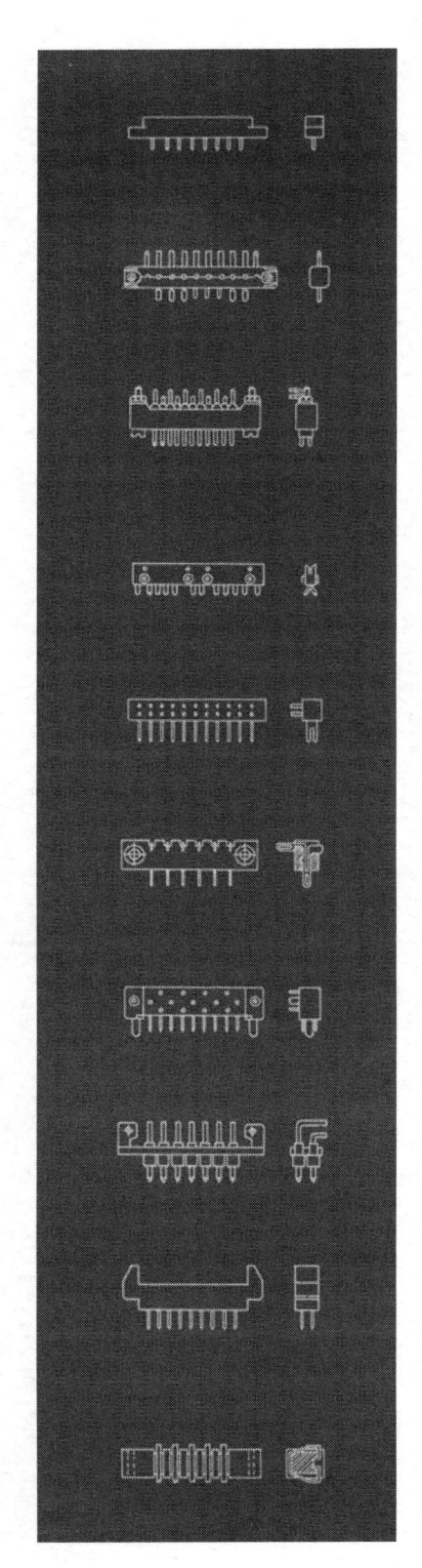

MA: *So, you're also saying that since no plasma can be destroyed, that the plasma of every living being that's ever existed on the planet is always in existence?*

BP: Well, that's for sure.

MA: *So is it possible to tune in to the plasma of any person in the past?*

BP: Right.

MA: *So in that sense then the ideas would be immortal or impermeable.*

BP: That's right. They are.

MA: *For instance, how would one tune in to the plasma of somebody from the past, if they didn't know how that particular entity existed?*

BP: Well, this is very interesting. Imagine that you sit at a radio dial, and you turn the dial. And you've been told in advance that if your dial is at a certain setting you're going to hear station number so-and-so. But if you continue this tuning process, you can come upon frequencies which are unidentified, unclassified, available to you because you have the sensitivity to correspond to this incoming sensitivity. Yes, this is very feasible.

HOLD ON TO YOUR HAT

MA: *So it's a matter of finding by random, is sort of what you're saying.*

BP: This is a random search.

MA: *Are you at all familiar with Konstantin Raudive?[5] The guy who put a bunch of thinkers in an anechoic chamber in West Germany and had all the people just sit there and meditate and think, not say anything. He turned on a high-quality tape recorder and recorded the sound of people thinking, and when they played the tape back, voices from history appear on the tape.*

BP: Yes! Well, this is what I just got through describing to you, or thought I described to you the existence of. In Albany forty people converged and I personally could feel the energy of all of them, and I could convert that energy to the new-new level that I referred to. Spirit-spirit level. Feeling-feeling level.

Yes, this is so. The energy of thinking is like the energy of living. Ever-powerful, ever-present. It is also universal, worldwide. Also cosmic, galactic. Ever-present in the universe.

While we're talking about energies, I think we should refer to Wilhelm Reich. His orgone is the energy of living, and it can easily be demonstrated by putting two straps on a tree, couple of feet apart, and connecting them to a galvanometer which will register an electrical current, which is being generated from the mere act of growing.

And this act of growing is in turn the act of living. With the result that our body, within three inches, has set up a field which Reich in his medical work was able to use to advantage. The area that did not give off energy was obviously diseased or defective or disabled, and could be felt by his traversing a body with his hands. He left his original work to be held for fifty years, hoping that in fifty years people would be in a position to understand what it was he was talking about. Meanwhile, of course, there are experimenters all over the world, not only trying, but actually duplicating some of the things he did, without having access to his original formulas.

---

5. Dr. Konstantin Raudive (1904-1974): Latvian theorist known for his work on Voice Phenomena–"Raudive Voices"

Interviews with Bern Porter

During his sojourn in jail, he was able to get out some of his formulas to a girlfriend, and these presumably are in the vault held by his daughter, not to be revealed until the fifty years are up. There are people who feel they can approximate what he was saying in his formulae and in his words. But it certainly would be better if they had the originals. Several attempts have been made to break the will and to have this original released for general use, on the argument that general use is not official. But so far the courts have ruled that it was his wish, and I'm a little vague at the moment as to the actual date when fifty years are up, but when it comes, I have a feeling we'll still be unprepared to really comprehend the full depth of what he was about.

Have you some other thoughts you would like to talk about?

MA: *Well I was going to continue on the plasma thing one step further, and since the Institute seems to be concerned with harnessable forms of energy, is it possible that the plasma that we talk about could be harnessed as a form of energy? Could it actually make things happen?*

BP: As far as we know, the answer is no. It cannot be harnessed in the sense you've been speaking of. Winds, sun, tides have been around for a long time and have been studied. There's a spot in France which generates power from incoming tide flow. There's a spot in Nova Scotia where it was tried for a time, with President Roosevelt, but this has been abandoned. The power interests pretty much control and discourage that kind of development. Again we're in the grip of the money thing, the power thing, who's in charge.

The best example of this, of course, is the automobile, and the fact that you get anywhere from eleven to twenty-two miles on a gallon of gasoline, when it is possible to get considerably more.

Another example of course is the steam car. I was in Tasmania at a point when the engineers would like very much to have had automobiles for their people. They had no oil or gas, a shortage of metal, but they could make a car of wood and they certainly could use water. It's interesting that through the simple act of using the post office, they could buy 1890 American patents of steam cars, and were able to duplicate those 1890 and '95 American patents for steam cars, in Tasmania, and actually propelled vehicles on less than a quart of water. Obviously the vehicle could not generate up to eighty mph in four seconds, which we seem to require our cars to do.

But back in the eighties, steam power for cars was not a novelty but an actuality, which we've managed to forget about—along with the barter system, the apprentice system, and other early devices.

What I'm trying to say is this: the old ideas still have validity, and have been submerged for one reason or another, mainly economics, the quest for money making. So as for your question, can this plasma be harnessed? In that sense, as far as we know, the answer is no. So that doesn't mean that we can't continue studying various angles of energy. The future of noise is considerable. The problem of course with all our past experiences is the waste factor. Even coal and gas have residues that have to be disposed of after burning takes place. So there's always a residue which constitutes a problem.

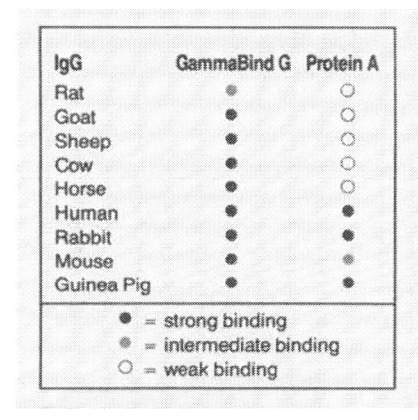

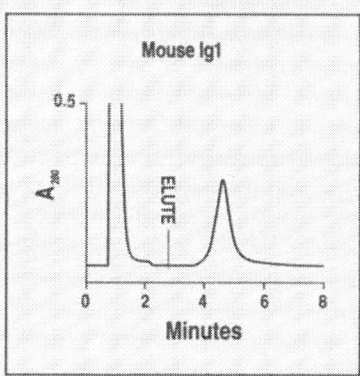

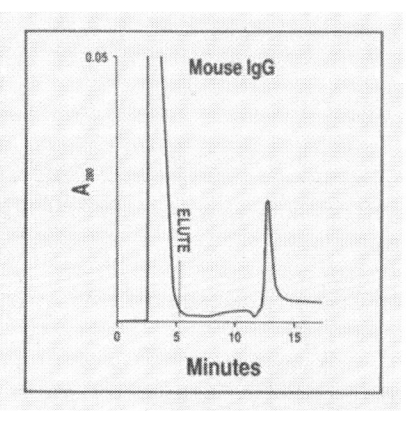

We've talked about your sneakers as an example of what we call at the Institute, *regressive technology*. Regressive technology is simply based on the making of money. It's its sole criterion for existence. Functionalism is set aside. You won't believe it, but I have a friend in Chicago who has a chair. Every time I go there, she cautions me not to sit in it. And I say, *Why not?* And she said, *It was made to look at.* So there are actually sofas and chairs that are distributed, advertised, wrapped and packaged with no intention whatsoever that they be used to sit on. I've seen decorators' stores and window shops in Chicago with a chair form: it has four legs, it has a back, it has a seat, it's a beautiful form in space, but all you can ever do with it is treat it as a form in space. It's not made to sit on.

So it's very interesting that most chairs that we do have don't really fit the dimensions of the average person in terms of his back, his hips and his legs. I've personally sat in chairs where it's not possible to take care of both the knees and my back.

So the sneaker is much out of line, the chair's out of line, and I'm sure you're experienced with razor blades–the single-edged razor blade, not just a support handle. Iin my time we used to get twenty shaves a blade, but now it's engineered so that you're lucky if you get two shaves out of it. And it's set up so that both the handle and the blade are intended by the maker to be thrown away. And you're also expected to buy these razorblades and handles in packages of six, and of course, what do you do with the residue of the throwaway handle? It doesn't disintegrate; the metal rusts. Now besides the razorblade and the sneakers, you're surely familiar with the pen. In the old days, you had a feather or a pointed quill, and you'd dip it in writing ink. A little later, there was a little lever on the side of a pen, where you'd suck up the ink into a rubber tube. Now, of course, both of those are gone, and you're expected to buy pens by the dozen, with the understanding that if you get two or three lines out of one of them you're lucky. Even here with you in the last two or three days, we've been struggling to find pens lying around, which cost money, and to get one that writes is something else again. And what becomes of those plastic shells? And the money-making deal goes on. And what's even more interesting to me, besides the fact that we used to have pens that would write and now we have pens in quantity that don't write, is that penmanship is no longer taught in school. When I went to school, penmanship was very much part of our training and experience. We were expected to write legibly, clearly and plainly. That is also gone, along with the pens in banks and post offices and those which we buy and cannot write with. We discard them, which the makers intend, because of course they're not recyclable or biodegradable. They're just thrown away, so that each person, in the course of a year, disposes of about six tons of weight ... just these metal, non-biodegradable plastic containers. 2,000 years from now, people digging around in the ruins of Lost America are apt to find some old razor blade handles and old non-writing plastic pens, all symbols of a very extravagant and wasteful society. Along with unfinished missile silos, atomic power plants and small-town industrial parks, whatever. So here we are, producing producing producing producing. An interesting thing is clothes. We're used to clothes being small, medium and large, and in terms of clothes for men and clothes for women. But there's such a thing as a unisex garment. The Japanese have pioneered this in the robe, which I brought to San Francisco at one point, and hoped to set up a group of American women to make them for American distribution. This robe is neither small, medium nor large, and neither male nor female. It's a robe, functional. Then you take the business of our food. First we cook it, then we can it, then we freeze it, then we thaw it out, then we cook it again. Whatever happens to the food itself? The other night you asked me if I would like to have a cheese sandwich, and I said, *Yes*, and you said *Do you want it toasted?* Well, the simple fact (and the Russians have demonstrated this) is that when you toast the bread, if there are any vitamins in it (and there certainly are), then you burn them or char them or distort them or destroy them by toasting. But nobody seems to be able to eat bread without the toasting. And the further complication on this cheese sandwich you were offering me was that you wanted to blend or heat the cheese. And then of course, we get back to society's social sandwich, we think of Dagwood, the comedian in the comic strip, his sandwich was six inches high, due to innumerable layers, and Big Macs from McDonald's–innumerable layers–and then we think, how can one have bread and cheese and lettuce and mayonnaise and bacon and whatever and whatever; bacon is not really a food, in terms of vitamin content. Meanwhile, our friends the astronauts were going up and living quite comfortably without pots and pans and spoons and knives and forks and tablecloths and stoves and refrigerators and chairs and candles. Just think, what if the common man on the street could have the same food in the same manner as the astronauts have it, namely just squeeze it out of a tube? There'd be no cooking. They also have a pill, about 3/4" square, just chew on the corners and it's the equivalent of a meal. Scientifically made up to contain all the ingredients that are necessary. Of course, the astronauts, it's true, are not active, but reclining throughout their space flight, and the food is generated and prepared at incredible cost, incredible, and it's not available to you or I. The nearest thing to it, and it's very crude, is prepared for hunters and fishermen. ...So here we are. We have everything and have nothing. Along with clothes, you think of beds. You have the single size, or the double size, the king size, the queen size, and each one of those sizes requires of course a mattress, mattress covers, sheets and blankets. When if the money angle were absent, we'd just have a standard bed. So it goes on. We consider packaging. It so happens that there's no country in the world that either excels at or approximates an American system of packaging and wrapping. You look at pieces of candy. They have two wrappings, and they're enclosed in boxes with further wrapping. And of course the automatic equipment that is devised is seemingly without end, and operates daily. Often these wrapping and packaging factories go twenty-four hours a day, seven days a week. And in this process, items are packaged and wrapped, whether they actually need it or not. And a staggering number of them do not. So we have a vast array of package designers, wrapping machine makers, and *treat and wrap* is the motto. Pre-cook and package is another one. Unpack in thirty seconds, warm it up and over, individually wrap the smallest, re-wrap the whole, there's money made where there was no opportunity before. And of course, we use concrete and reinforcements, and it takes fifty years for the use of steel in ships, and steel in skyscrapers. And more unfortunately, of course, with what the Institute refers to as regressive technology, we have forbidden technology. In the case of the astronauts food, as an example, or as in the case of a car which goes only twelve miles. And so it goes on and on. And as well as regressive technology and forbidden technology, we have *secret* technology. This is where patents are indicated and registered, but if they were put into practice they would revolutionize current practice. So these are filed away and never heard from again. So what does one do? What chance, when you have regressed, forgotten, secret and crackpot technologies ... where's the individual going to set in? You also know, we have Jell-O available now, something like thirty-three flavors, and in my time there were only two, or three, and the staggering number of whiskey brands, and of course all of this takes up labeling and wrapping, so it's not a very pleasant thing to contemplate. And of course, our American-made appliances, machines and tools—any mechanical or electronic equipment that you could name— suffers something like 93% depreciation loss twenty-four hours after you purchase it. Made to break down, not made to last, in perpetual need of repair. And for most of 'em it's cheaper to throw out than to fix; all of 'em have replacement parts that aren't

HOLD ON TO YOUR HAT

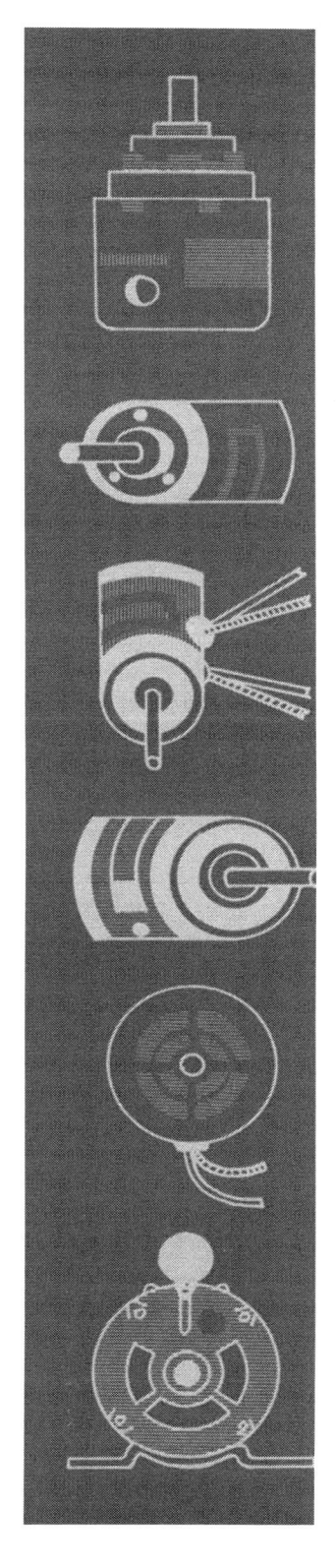

easy to obtain, or are more often non-existent. Some are obsolete in all respects in six months. And of course they consume more energy in production than will ever be regained, by any known recovery means. We've wasted countless irreplaceable materials and resources, and in general, they're not essentially needed. Well, in this calculated set-up, the junk men are the king rulers. The bankers and the tax collectors are the associates. Well, I suppose you may ask me, is there any way out of all this? We come back to the principles of Simplistics that I started to mention at the beginning. We would prefer to complicate what is complicated, we prefer not to make simple what is complicated. And I spoke of the division of the day, in which one could divide his personal life from six AM until noon, and say, *In this six-hour segment I'm going to resolve and take care of all the problems, and then I'm not going to worry about anything else except for these six hours*, and at noon, we should forget about the past six hours, and confront the oncoming, from noon until six at night. Divide the twenty-four hour day into six-hour periods. Each period we use to the fullest advantage. I have some other examples of what goes on, and I'm sure that you could think of others that you could mention. In my case in Belfast, I use sixty gallons of water a quarter, and I have to pay for one-hundred gallons. One-hundred gallons is the standard set, and I don't conform to that, but I have to pay. And the same way, while I'm using sixty gallons of water and have to pay for one-hundred, in the town where I live, there's approximately fourteen other people who're also doing the same. And the same with electricity. I use sixty kilowatts of electricity and have to pay for one-hundred. And the same way with premium insurances. We think we have insurance to pay for damages, but the insurance company thinks we have insurance to pay them premiums, and just as our damages exceed a certain number, our insurance is cancelled. So we have these disparities of a very disturbing kind. In any system you want to mention, there's this overlap, this duplication, this waste, waste, and in all of these cases they're burning up energy. And energy in turn must be created to supply the need, and here we are, just waste, waste, waste. Well, so what?

MA: *Maybe I'll just ask some questions, and we'll see what kind of response the Institute might have to these questions. Have you ever done much exploration in sentience of plants, intelligence of plants?*

BP: Well, this is a virgin field which has been highly overlooked. It just so happens that plants are living, plants have plasmas, plants are talking and communicating with one another. And they certainly are generating energy and they certainly can be converted to alcohols, and the alcohols can be used as a fuel, and in turn the sensitivities which they give off have yet to be fully understood and explored. As far as we're concerned, it's a virgin field. In many primitive cultures, including the native peoples of Maine, they're more in tune with that than we are. They took goldenrod and other plants and made dyes and beverages that had functional uses, and in addition, they understood what a living plant was. One of the great horrors of our culture is the acid rain and pollution of the worst kind, a staggering number of botanical specimens are becoming extinct. An incredible number of specimens have already disappeared. We're concerned about the bald eagle and the bison and buffalo and whatever, but we've never cared one way or another whether they are gone—not only gone, but in the long sense gone. Yes, the whole field of plants, the whole plant world has been highly misunderstood, and highly worthy of continued research in the sense that you speak of.

MA: *Do you have anyone on the Institute on that?*

BP: Yes, two people. It's interesting, a man in New Zealand is concerned about this, because, in Tasmania, where there are no minerals or oil and metal for cars ... in New Zealand they have a need for a great many things which could come to them through the plant life that grows there. I went to Tasmania to help them make paper out of hard wood, because they have tremendous growths of hard wood, and paper normally throughout the world is made of soft wood, and so there's an engineering and physical problem concerned. What I'm saying is each environment has qualities and substances that are characteristic to that environment, that can be used to support the lifestyle of those that live there. It's just a matter of research and development. What I'm saying is that the field for research is wide open, that's still untouched, things we do know are under continual re-examination to find out if they are as valid and as true as we think they are. I was interested in these pictures you've just brought from Detroit, which show debris, or what you call junk, used for decoration. And recycling, in an interesting way, an aesthetic way ... putting pieces together, which comprise various colors in form and space. This is a sensitivity ... I suppose the creators are called *junk artists*; these are people who are in tune, certainly, with what's around, and are aware of the waste, and the recyclable possibilities are certainly anything and everything.

MA: *Yeah, I would suppose that in a culture that's in ruin, that's about all that you're left with—the garbage.*

BP: Well, as I say, an individual is producing something like six tons a year of what is called garbage, and we've reached a point where we now suddenly are obliged to separate our garbage and glass and metal and paper; we're slowly coming to realize what a garbage-producing culture we have, and that in effect we're burying ourselves in our own productivity. That's what I was referring to when I said that we have thirty-three flavors of Jell-O and 108 kinds of whiskey and sixty-four kinds of soap. We have this tremendous multiplicity, but at the same time, where are you going to put it? How are you going to re-use it? So the artists, again, are pointing a way. My whole theory of found expression concerns the simple act of going to the post office and taking out of the waste basket of the public room of the post office anything and everything, carrying it home and going through it and finding what I call gems, verbal gems, visual gems, cutting them out with scissors and re-pasting them. So I feel that as an artist, I'm pointing a way or ways to bring back some of these things to practical use. The whole world of visual poetry rests on this recycling of visual imagery. Which reminds me of an attempt in Maine to revolutionize and overthrow the haiku, in development in Japan some 8,000 years. Is it not possible in our tremendous verbiage which is thrown on the side of the wall every hour, to find a few gems which could be the equivalent in terms of syllable numbers and line numbers, to be an American form of a haiku? I think this is a valid approach, and worthy of continued attention. And again, it's interesting to me that it's the artist who's approaching this subject, through the world of clouds, through the world of vision, through the worlds of re-creating, getting on, getting forward. Of all the people, only the artist seems to have the sensitivity or the concern for things in space, color in space, materials for re-use.

MA: *Some people would argue that found art is plagiarism.*

BP: Well, this is quite valid if the things that are being re-used are owned by someone in the legal sense of being registered and copyrighted. On the other hand, in my work I go out of my way to make sure that it's not, but even then I'm accused of thievery. I take the position that when one finds something, no matter who or where, and reshapes it even to the point of turning it around, or turning it wrong-side-out, or turning it upside down or seeing it from a different angle or imposing something on it, then it becomes a new form through the act of re-creation, and is not plagiarism. Plagiarism is just outright copying, duplicating, but if you take something and use it as a source, bring to it variance, to re-express, then I think the word *plagiarism* is pretty much overdone. Actually, it's not possible to own anything. Everything belongs to everybody.

MA: *I think part of the reason that the whole plagiarism thing exists is the recent ability of technology to reproduce accurately. Anything - audio, image, ideas, brainwaves . . .*

BP: Well, technology is always open to question. Is it saving us, is it actually helping us, or is it deteriorating us in ways that we're just not aware of yet? It may take us quite a few years to find out that what we thought was so helpful is not. We see drugs

HOLD ON TO YOUR HAT

being withdrawn, we see automobile makers withdrawing models, we see in Europe people tossing over forty years of Marx and Lenin, we see continual recycling, re-examination going on, and this is a very healthy thing. And this in turn brings us back to the premise of the Institute. We must see all things in a new way, in a fresh way. I guess I mentioned the case of Fuller and his globe. The globe was cut into triangles and laid flat upon the floor. So we could go on now. We've mentioned this day a number of things here, all of which could take up a concern if we were interested in efficiency. And if we're interested also in conserving energy, all these gadgets that we're talking about burn up incredible volumes of energy, electrical and otherwise. Which reminds me, I went to a group who said *Well, we have to conserve energy.* So I said to them, *Well, are you people willing to go home and turn off some lights?* And I told them that in my cellar at the Institute, there are six light bulbs in the basement on one switch at the top of the stairs. It would be possible just to unscrew five light bulbs, turn the switch. Then only one necessary bulb comes on, then go down and screw needed others back in. And then I was talking about neon lights and going door to door.... It's very interesting that in Sweden, they figured out that they needed three new atomic power plants in so many years. They also figured out that if they went door-to-door, out of a list of something like 104 electric gadgets, if they just knocked on the door and said, *Do you have any of these 104, and if you do, the fine for owning this or that is so much, and the fine for owning it is greater than the cost of the thing. It'd be cheaper for you to just give it to us.* So they just clean out houses of all excessive gadgetry that consume electricity, and finally ration electricity to only four hours a day. So we have these incredible gadgets: meat cutters, clocks, shavers, shoe shiners, which are consuming electricity.

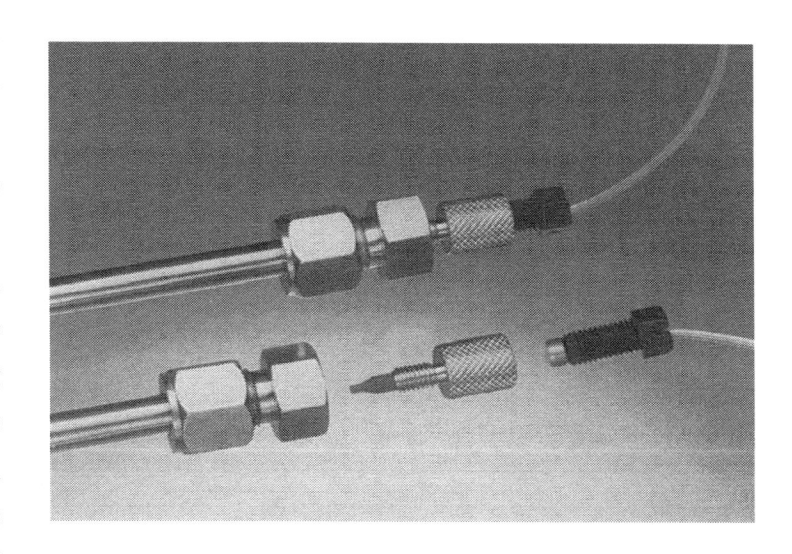

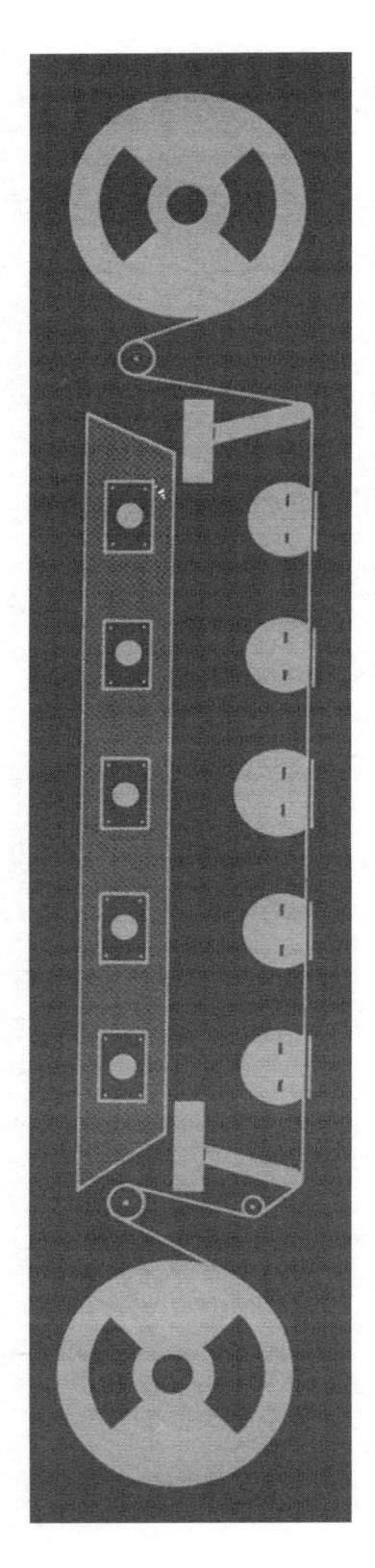

As a result of this act in Sweden, the story is that they demonstrated that they did not need new power plants in six years. All they needed to do was stop the use of their present electricity by eliminating an incredible number of gadgets which use it. And this again is a case of making things simple, and also doing things in a way that's obvious and reasonable. Our simple Law of Simplistics is to stop making things complicated. Let's see how we can simplify them, make less and less. We can make more out of less. So this is an unending process, also very discouraging because the chances of having a steam car on my own street out here tomorrow … a cube, half an inch square, to eat three times tomorrow … eliminate pots and pans and refrigerators and stores and candleholders and whatever. All these things that I've been talking about are not only far out, but unobtainable with the way things are set up now. We refuse to change, to alter or even to adapt. So we just bungle along, like we've always been doing, I guess.

MA: *You've been talking about the conservation and ecology of energy, but what about the conservation and ecology of mind power and imagination?*

BP: Well, this again—like the business of the plants, what do plants have to offer us—has to be explored. Certainly the act of creativity is a powerful generator of a number of forces and talents, some of which we just don't understand. There are people who take a pencil and make a few whirligigs on a piece of paper; these can have meaning in theoretical physics and music composition and poetry. These topics have yet to be explored and understood. We have names for them ... we have so-and-so at some corner of some remote university, working on it, but the university will not allow it to be publicized. An interesting example of that is the so-called Cold Fusion. These two gentlemen produced some electricity at a university, and everybody got excited, and the reason they got excited was that they continue to use the word *fusion*. What happened was *not* fusion. It was an effect which they have yet to understand and to name. And this is a part of our horror—we have this tradition of using certain words for certain things, and when something new comes along, we don't know what to do with it except put it in a pigeonhole of outdated use. So we're stuck all the time. It's like you and your cheese sandwich—you, by convention, insisted that it be toasted, and the cheese be blended. And all I want from you is a piece of bread in my left hand, and a piece of cheese in my right hand! My senses were to smell the cheese when I masticated, and the same with the bread. I was to both taste the two elements, and smell the two elements and enjoy the flavor of the two without this intermediary business of mayonnaise and toasting and layers of ingredients. So here we are ... the great staggering number of things the mind ... what the mind does when it thinks, when it creates. And can that power be translated? The field is wide open.

MA: *Most institutes publish newsletters. What about your institute?*

BP: Well, our memorandums go between us. We do not give them to the public until such time as all of us are convinced that what we have is functional, doesn't use more energy than is required to produce, that it is obvious, that it is reasonable, is sensible, and this is merely available in announcements. We do not have a newsletter in the form that you use. And if you're familiar with newsletters, they're in general deceitful and incomplete, and usually out of date between the time they're written and the time they're received. So they constitute a waste of time, energy and paper. So we do not have newsletters.

MA: *Have there been any findings that you have released to the public?*

BP: Yes. And all of them are based on this business of conserving energy, this business of being reasonable and simple and uncomplicated. I spoke of the breakdown of the forty-hour week when we started talking. And I spoke to you about the necessity of discovery of the way in which people walk, and now how people sit, the use of beds and clothes. All of these are basic, fundamental things which we're endeavoring to obtain. It's a very discouraging kind of operation. Take the case of one of our members who grinds his own lenses. Who has time to grind a lens? (Or even the inclination, the knowledge or the ability?) We have another artist who mixes dry color pigments with beeswax and assorted waxes, and egg yolks and varnishes, and uses techniques of preserving color and emphasizing and augmenting color. So these are continual processes that go on, and newsletters would impede them.

A general announcement that we make is that people come to Belfast between May 1 and November 1 and bring themselves, a tent and a sleeping bag and sleep under the stars on the open platforms or in their tents, and bring hiking shoes and walking shoes, and spend time hiking and walking everyday, and swimming and thinking about some aspect that can be simplified. Some aspect that can be made reasonable, some aspect that's currently unknown, like plant minds, all of which are around us. To see old things in a new way, that's our main message. And coupled with that, of course, is the development of our senses, to use as many of them as possible, at all levels at all times. This is quite an act in itself. And there're no exercises or forms for doing these; it varies, individual by individual. Some people feel like they can't get outside themselves unless they have drugs, or if they smoke, or have wine, or if they dance or exercise or all this. These are all valid, but only to a limited degree, and basically are not necessary. There are many cultures in the world where they have exercises for stimulating the body or maintaining the body or developing muscle power or any number of purposes. The trick is to combine all of these in a way that is efficient for each individual lifestyle and individual desire. In the final analysis, there are no words that describe this. After the words give out, in theoretical physics we invent symbols, and we juggle symbols. And of course after the symbols give out mathematics breaks down, and what do we have left? Well, we have what we call the feeling thing again. Gertrude Stein, and a rose is a rose.... Well, in addition to the rose, is the roseness of a rose, and there is the spirit of the rose, and the rose mind, and the rose's perfume, and the rose's use in cosmetics, and the variation of the variation, and the newness of the new. Words just break down and don't describe what I'm talking to you about. And this is very unfortunate.

MA: *I think sometimes this duality of simplicity and complexity is a problem with language; that sometimes it isn't even a matter of making a situation more simplistic, but the language you have for describing that situation. We have a word that we've coined, simplexity, to describe that concept, the idea that often, a situation can become more or less simple or complex depending on how you describe it and the emotional state that you're in when you describe it, and the whole aura of the moment.*

BP: That's correct. It varies from individual to individual, the state of his digestion, the state of his rest, the state of his diet, the state of his hormones, his longevity, his lifestyle or whatever, and then of course all of this is taking place in the time frame. So it's not possible, actually, to get a grip on anything.

MA: *So do you have any people at the Institute who work specifically with the problem of—well, it's almost like a Wittgensteinian problem in philosophy—the language itself. Learning what kind of language you need to couch problems in...*

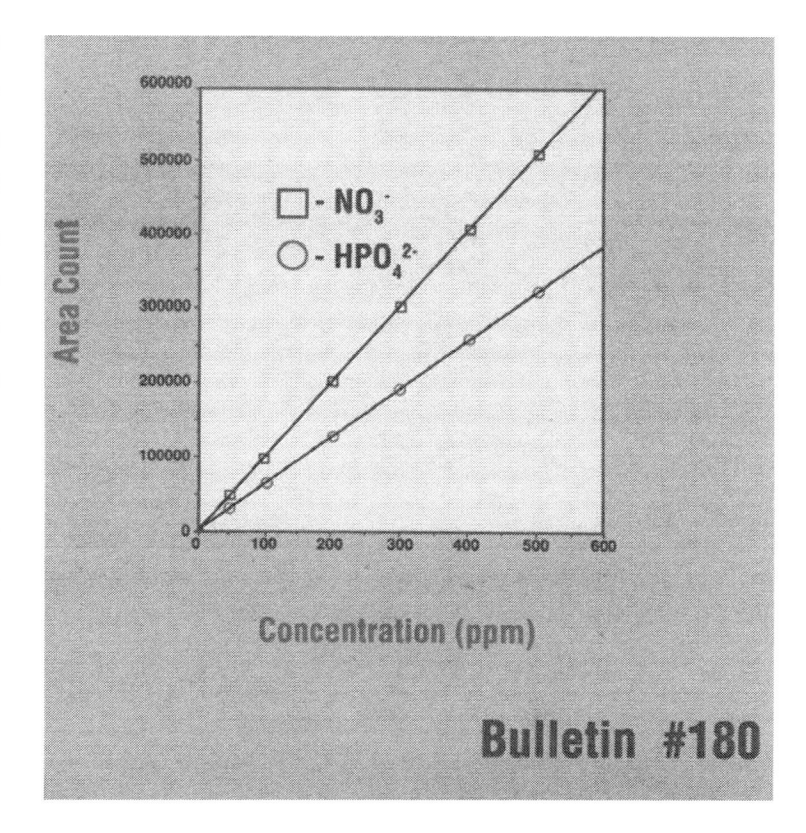

BP: Yes, well I just got through saying that language is a barrier, words are a barrier, it's amazing we can talk at all, and that we almost have to throw out language completely and get back to the action of frequencies of vibration of individual plasmas again. The things that we talk about that really matter, you cannot express in words. Or language.

MA: *Do you spend much time thinking about the future? Maybe trying to put yourself 1,000 years into the future?*

BP: As I pointed out, in our concept of physics, we're merely in the moment, progressing from moment to moment. In this there's no future or past, all there is the present, all we can say is that the present will proceed through time, time will push forward. We will come to some condition for which you now use this unfortunate word *future*. There are many words like that ... I spoke of the fusion, the cold fusion experiment, well, it's not fusion. We're stuck with traditional uses, and one of 'em is this word *future*. Well, what is meant by *future*? See, we have four dimensions, the fourth dimension being time, and it's this time factor which will take care of this thing you call *future*, and it's too bad that there is such a word. What does it mean? You want to measure it and say, this is 1989, what's it going to be in 2000? 2010? 4010? Well, we cannot predict. We can merely proceed with time.

MA: *Well, is the Bern Porter of 1930 surprised by what the Bern Porter of 1990 sees?*

BP: Not particularly, no, because things have basically not changed. We're still wasting, we're still being inefficient and incompetent, we're still creating waste, we're still burning up a slew. No, I think the things I had in the first three months of my life were completely contained and included in anything that's going on in '89 with me. See, I have to get you used to the idea of things happening simultaneously. This is very difficult to conceive of. A staggering number of things take place at once in this time frame that I speak of, and the time frame of the present (*present* again being a very unfortunate word), and it's hard to understand this word *simultaneous* and this time frame. You speak about word changes, these outward changes that are again followed up with language. Language is very unfortunate, and the use of the word *democracy*, a word like *politics*, words like *peace* and *future*—these are incredible barriers.

HOLD ON TO YOUR HAT

All of the characteristics of your life were established instantaneously, and these characteristics that were established, formed and set up by the design of the ingredients in your father's and your mother's *semens*, as it were, can never be changed; they're with you for the rest of your life. And the church and the school and your parents and whatever you go through cannot change this original pattern, this original design. And very few people seem to realize or be aware of this. So they go to school or they join some organization or they take exercises or they do this to change themselves, but *this is not possible to do*. The plasma has been generated, all of its characteristics have been established.

Then of course no two are alike. That's one of the features of nature. Then you ask me is there any change, do I find any change, the answer is no. There's no change. I have to admit that I am suffering from what most people call old age. My hair has turned white and I can't concentrate, I can't remember; but I had those characteristics at birth. And I can't complain.

It's very interesting that this *old age* thing is a matter of physical deterioration, and the exciting part of this is that mentality and the function of mind … while I can't remember, I still, as a result of this passage through time, am entering into outer layers of understanding and of knowing, and perceiving, comprehending and feeling. But all of those were with me at the time of inception. It was understood at the time of conception that these things would remain unchanged but would *fulfill* themselves, and increase in richness and in value, and substance, but would remain unchanged.

So while outwardly I'm confronted with what you call old age, in appearance or whatever, I'm still not only young but extremely young in the areas that matter. That is, I can see and feel new things, new shapes and new forms, every moment that passes. So it's not possible, in the system that I'm speaking of, to grow old. The container of my plasma certainly is growing old, there's no question about that, but the plasma itself has been established according to a pattern and design that remains intact. You can say, *Well, this is a miraculous form*, and I have to admit it is. It's beyond words. You can't even describe it. Only roughly.

Interesting thing to me about physicists is that they are, in my understanding, the most religious, if you want to use the word religious, they are the most religious of all of the sects. Any sect that you might mention has rituals. Even burning of candles or of incense, or songs or prayers, or Bibles or standing and kneeling and whatever and whatever. But physicists don't need any of those external decorations. They are continually in the presence of all of that, and don't need this so-called religious atmosphere to instill it or to bring it about. These are all artificial things. People go to church to light a candle or to see the incense or to be hypnotized by words, prayers or songs. To get into what they call a religious fervor. Well the religions and the religious are very unfortunate, and physicists have all that, because they experience it daily, twenty-four hours a day. They don't need this external decoration and ritual, they don't need drugs or stimulants.

So maybe we're a very special kind of people, I don't know. But I don't think so. Basically all people are physicists in the sense that they are exploring *something*. Even if it's a simple matter of how to make more money, or how to plant the garden next spring. I've always been impressed with the fact of the design, of an original pattern, and the fact that it remains unchanged in the course of expressing itself. And all of these actions, as I say, take place in what we call simultaneity, and this is not easy to describe. How do you describe something that is both living and dying, that is both expanding and contracting, that is both stagnant and moving, that is both active and passive?

You see we're used to having a thin line of demarcation between those, when they're one and the same thing. People ask me *Well, what about physics and art?* Well, they're one and the same thing. The creative process that goes on in each is one and the same, and the result is one and the same. See, we're inclined to divide things by labels, and by classifications and by words and by definitions and rules and regulations and codes. After you get through all that, what have you got left? Well, mostly nothing. So we make a morass out of everything.

MA: *Some people think that chaos is predictable.*

BP: Yes, chaos is one of these states which accompanies every phenomenon; you have to realize that in nature there's the positive and the negative. And that these are ever present as good and evil, and as chaos and static, and there's unrest and rest.

HOLD ON TO YOUR HAT

So again we're back to the simultaneous thing that I'm speaking of, and we're also back to the time frame. And also the simple fact that two observers don't necessarily see the same thing at the same time. I'm moving this glass with my foot on the floor, and from my point of view I can see certain things which you sitting there two feet away may not see. You may be seeing some things in this action I don't see. Just because, you're looking at it at a different angle than I am, or your eyes may be in the light, or maybe for you a little stronger, or you may see some shadows that I don't see. And so it's impossible for you and I to describe what is taking place here. And coincident with this seeing force is the condition of our bodies.

MA: *It seems that traditional religions— Zen Buddhism, Tibetan Buddhism, Hinduism— are really primitive forms of physics that were established thousands of years ago; once again the thing that separates it is the language that's used, but it's really talking about much the same ...*

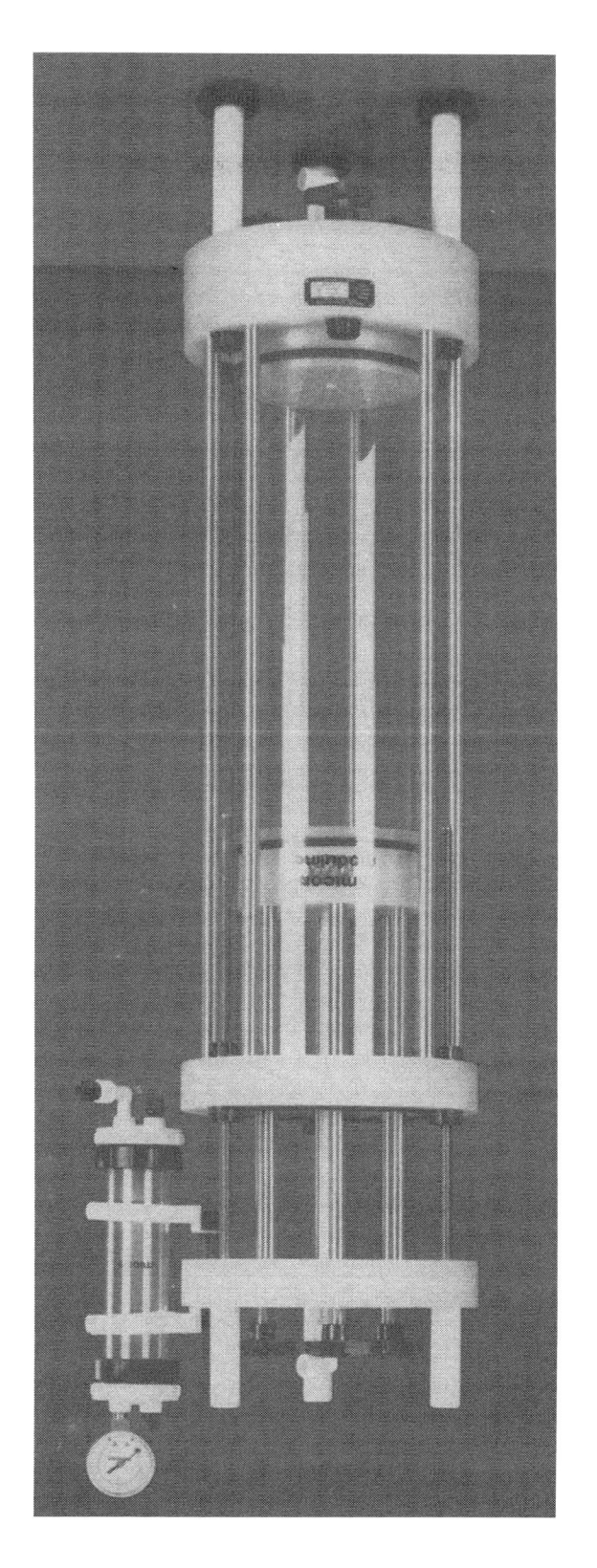

BP: That is very true, I appreciate your bringing it up. All of these groups that you mentioned are talking about one and the same thing, they're just using different approaches, different words. One would use incense, the other would use candles, others sit under a tree for 40 years; they're all talking about one and the same thing. They're just approaching it from a different angle. You're sitting there seeing what I'm doing, I'm seeing what I'm doing but what I'm doing is really indescribable.

MA: *Aren't they all essentially a model of the universe?*

BP: Yes, yes they are. Yes. Yes, and what is the universe? Well, it is a combination mass, going through time. Time's going through it. And it's doing so within a pattern, right? A pattern of the greatest efficiency.

When I was in Albany some people were asking me about my experience in Hiroshima, where a boy comes up to me and has in his hand a mass about the size of a golf ball. Well as I look at it, it takes me quite a while to look at it, but the more I look at it, it slowly comes over me that this mass in his hand now the size of a golf ball used to be an empty milk bottle. The fusion, the heat and the pressure of an atomic explosion caused this empty milk bottle to close in on itself, and it melted, fused, re-melted, refused and assumed the simplest possible shape for itself, which turns out to be a sphere. This is a natural phenomenon.

So looking at this glass I'm reminded of this boy in the street in Hiroshima, how this glass here on my foot can close in on itself, the water can evaporate in the process, and literally reduces, you would think to just a flat pancake-like form. But, no. If the pressure and the temperature's high enough, it doesn't have time to flatten, it has only time for closing in on itself.

Well, so the observer, the condition of the observer is the determinant factor and he's limited by words, and words are not enough. They're deceptive. Not only deceptive but unfortunate.

So you can say well, what do you do when you think? Well, first of all I believe you bring into play just about everything you ever experienced, no matter what you're thinking about. It's an instantaneous summation of everything that you ever experienced, read, felt, heard or whatever. And in this thinking you're trying to resolve and resift and restate all this, what you've gone through, what you've heard and know. And you're trying to come up with what we call an answer, or a solution. And you're doing this in terms of your environment, your particular situation—financial, emotional, and otherwise. So a simple act like thinking suddenly becomes very complex and it's probably not completely understood, even yet, what goes on when you think.

An even more interesting question is, who is the first guy in civilization to use the word *think*? Or to use any word? I'm sitting on a chair, who was the first guy who said *chair*? and the word *chair* can be translated into something like 25,000 languages that we have, and all of them come back to this form. Well who was the first guy to say it, what was he talking about? And this is even more interesting, was he talking about something that had four legs, a seat and a back, or what is a chair? But more important, who was the first person to use this word, or its equivalent in any language? Well, of course, I don't know if we can say.

Well, we've gone over a long range of things here, in this discussion. And all of them I think come back to some of the basic principles we've devised, namely: look at old things in a new way, increase and preserve what senses you have, and be aware. The condition of being aware, of course, is the use of the senses to the full.

So I guess we can conclude our discussion for the time being, maybe I'll continue writing this out, which happens to be rather formal. And it's true I can gather some illustrations of one kind or another, some of the things on the Institute grounds, and eventually we can get a printed form. But until such time as we do have a printed form, this conversation that we've had by tape will have to suffice, and may even be in many ways better than the written printed form. And I'll continue working on it, I have some fifteen pages so far, and I have more to go.

So we'll call it the end, with the understanding that there is no end.

MA: *Thanks, Bern Porter.*

Interviews with Bern Porter

Photographs by Steve Random

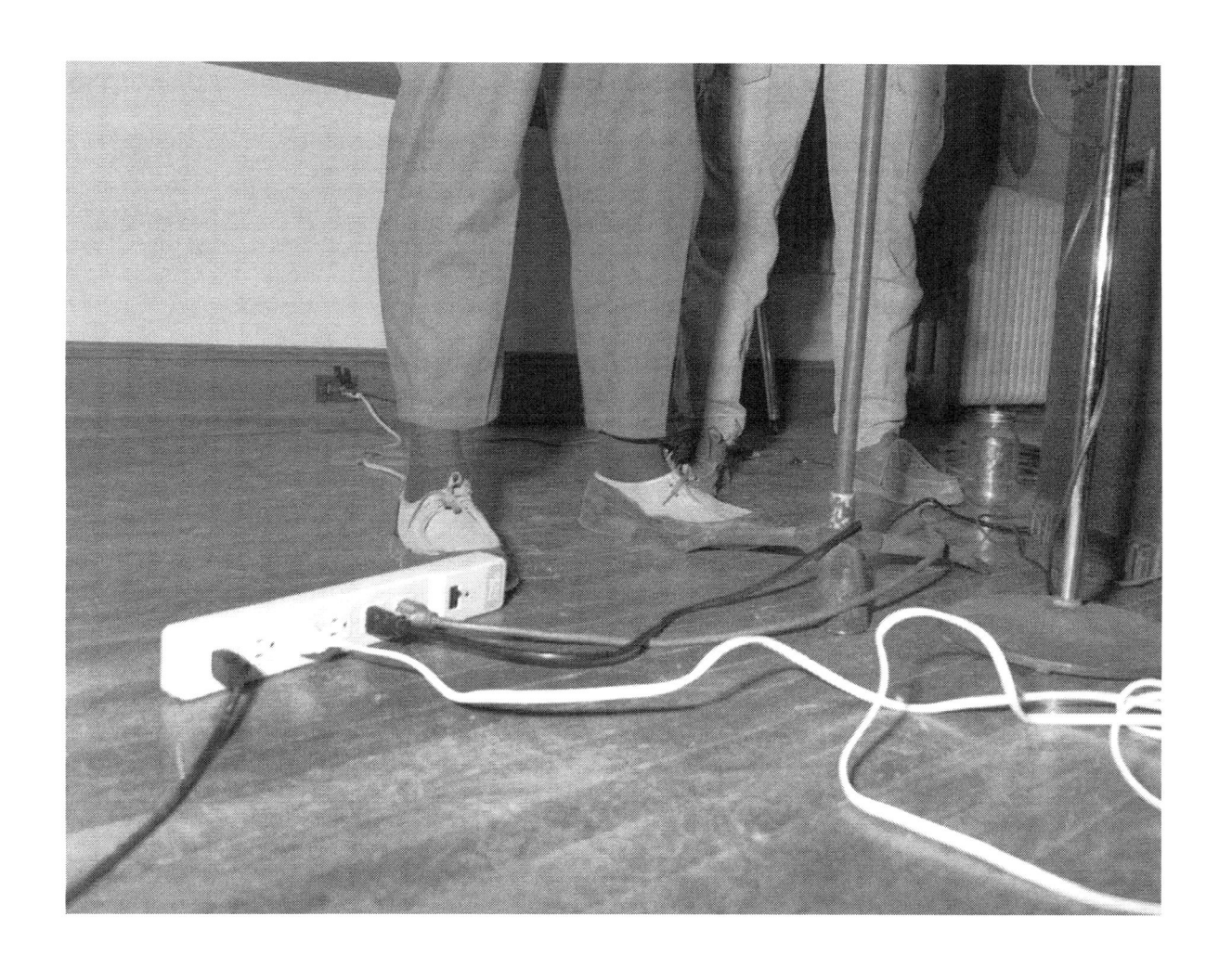

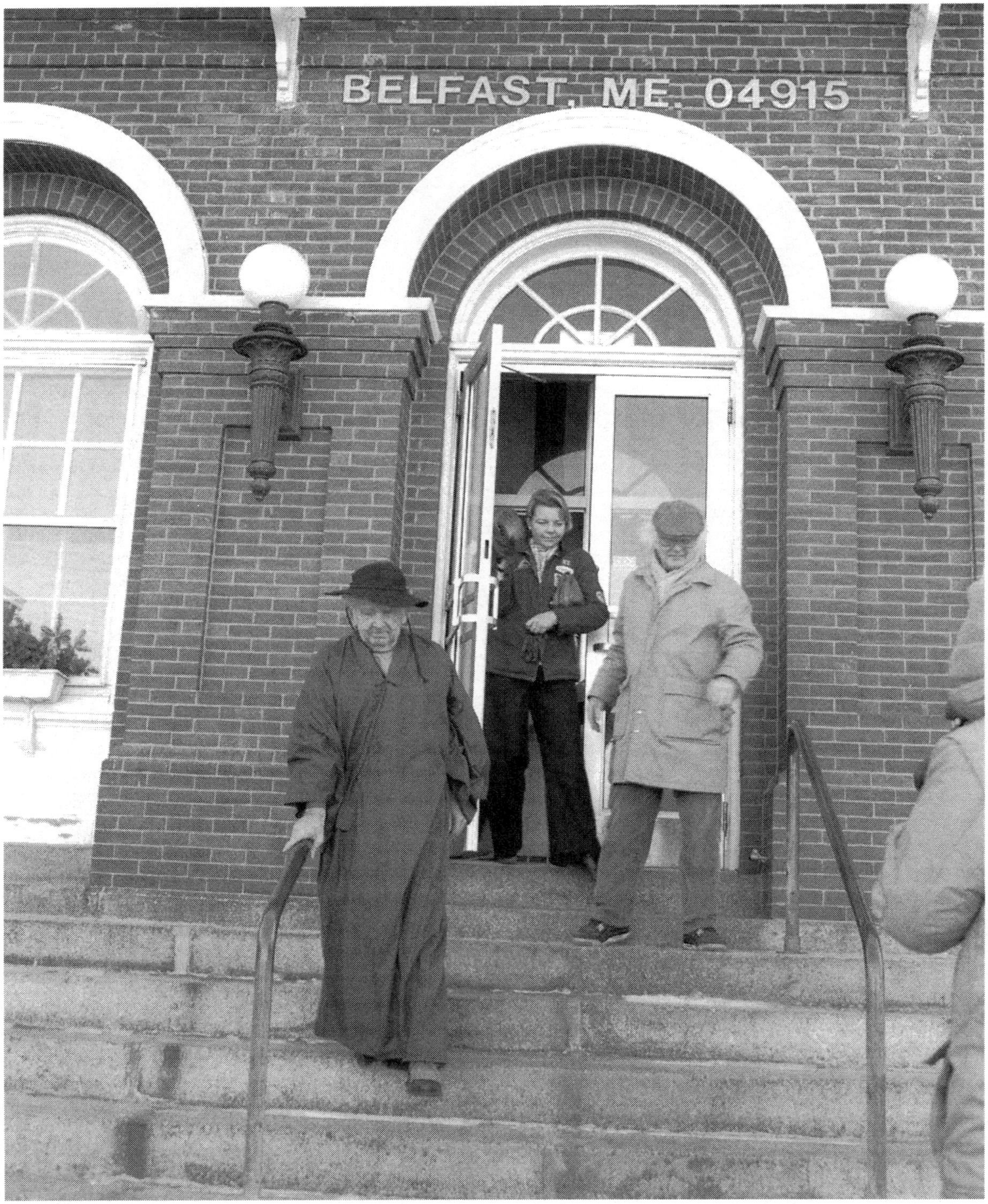
BELFAST, ME. 04915

HOLD ON TO YOUR HAT

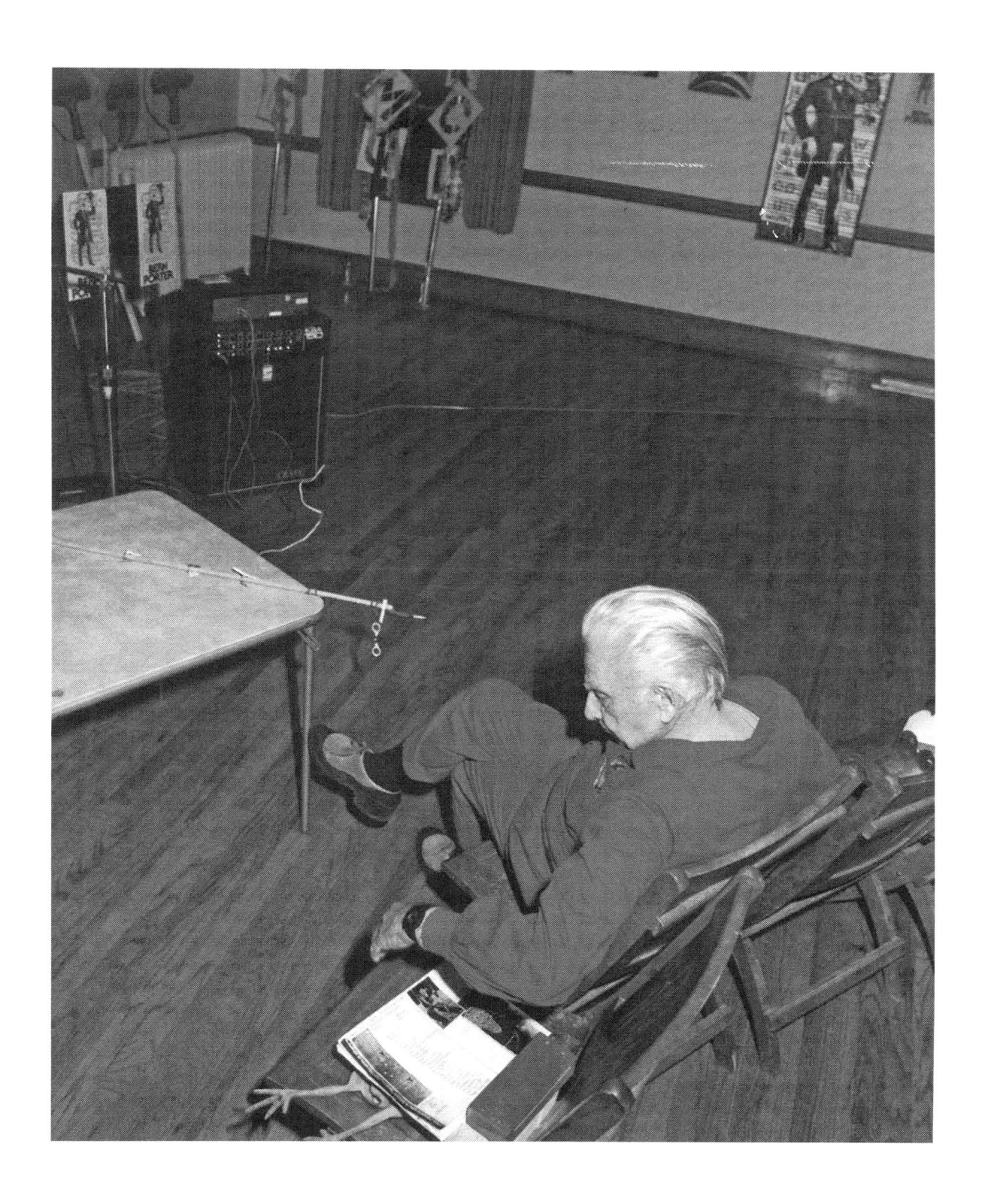

Interviews with Bern Porter

HOLD ON TO YOUR HAT

Interviews with Bern Porter

HOLD ON TO YOUR HAT

Interviews with Bern Porter

HOLD ON TO YOUR HAT

Interviews with Bern Porter

HOLD ON TO YOUR HAT

Interviews with Bern Porter

Questions for Bern re:
The Institute of Advanced Thinking
mIEKAL aND, Elizabeth Was & Ben Meyers
December 1993

mIEKAL aND: *How many dimensions are there in your world paradigm? We know about time, space, gravity, magnetism, and electricity.*

Bern Porter: 24 basic dimensions minimum
40 basic dimensions maximum
used daily in Institute investigations
radiation—anti radiation
energy—anti energy
force—anti force
pressure—anti pressure
orgone—anti orgone
plasma—anti plasma
time—anti time
space—anti space
gravity—anti gravity
magnetism—anti magnetism
electricity—anti electricity
wind—anti wind
tide—anti tide
sun—anti sun
living—anti living
dead—anti dead
seeing—anti seeing
hearing—anti hearing
feeling—anti feeling
sensing—anti sensing
smelling—anti smelling
plus others we are working on

Interviews with Bern Porter

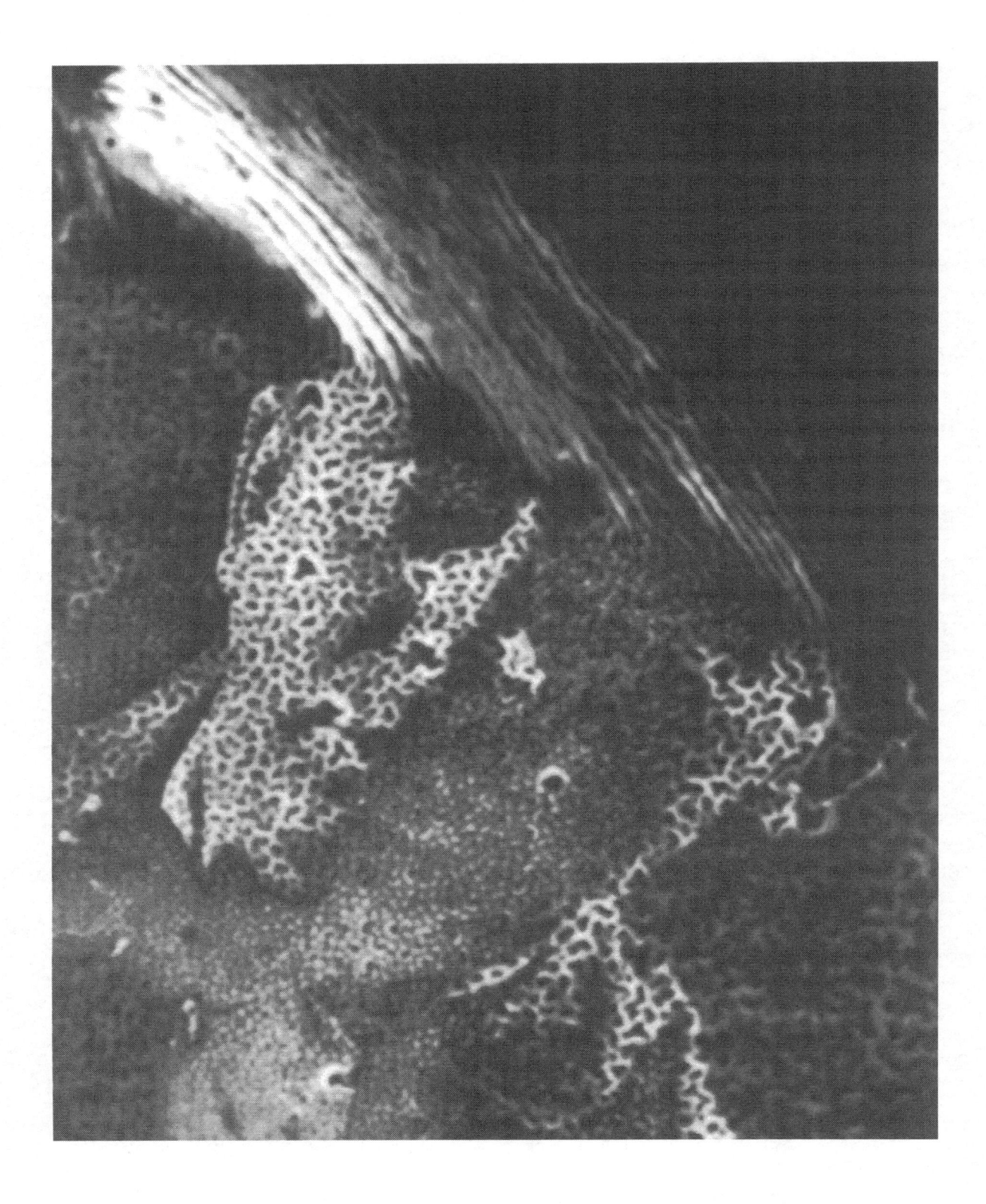

HOLD ON TO YOUR HAT

or
to every positive
there is a negative
to every prime a secondary
to every effect an anti effect
to every reality an anti reality

MA: *What are the most influential books published regarding the new sciences?*

BP: 1. all titles by Buckminster Fuller
2. all titles by Wilhelm Reich
3. all titles by Stephen Hawking
4. all titles by Richard Feynman

MA: *We have known since Tesla that the atmosphere surrounding the earth contains enough energy to fulfill all the needs of the planetary population many times what it is now. It is understandable why the forces that be have covered it up, but why haven't the underground intelligences made black market devices available?*

BP: fear of repression
fear of being found out
fear of being branded a crackpot
fear of being hounded by commercial forces
fear of being sued by commercial powers
fear of government regulations
fear of environmental codes
an overwhelming pressure
to maintain status quo
coupled with universal ignorance
and lack of forward progress
and adventuresomeness
and spirit

Interviews with Bern Porter

MA: *What is your experience with UFOs? Have members of the Institute been in contact?*

BP: None of us have ever experienced such. Whenever we hear of anyone who claims to have we do all possible to see and interview the person. In general we find these people—some ten by now—deranged publicity seekers. However we do feel there are intelligences on other planets equal and superior to our own who have not yet visited us by mechanical devices.

MA: *Can you make any Institute papers available for publication in this book?*

BP: Sadly no. Each member, without either consent or knowledge of Belfast headquarters, can and often does solicit both use and public promotion of their ideas and explorations in their general area of location and subject.

MA: *In the year 2007, Wilhelm Reich's papers will be made available to the public. Do you think it was wise to hold back information that might conceivably avert many world disasters?*

BP: I and our members have for years been both saddened and much frustrated, even hampered to a very high degree, by this deliberate legal and rightful withholding. Slowly we, all of us, are forced to believe—hard as it seems—to agree maybe we are not yet ready and sadly may still not be in 2007.

Elizabeth Was: *I get the feeling that you function as a director. Who will carry on the task of keeping everyone working on projects?*

BP: This remains a very serious problem for which I now have no answer, only the hope that someone will show up to carry on. I have no children or near relatives so oriented or able. At my passing the entire property, contents, and reports go to the President and Trustees of Colby College, Waterville, Maine, whose Special Collections Department at their Miller Library permanently houses and catalogs more than four thousand items of my donation since 1974 in what is called the Bern

HOLD ON TO YOUR HAT

Porter Collection of Contemporary Letters. I had hoped and continue to hope that Colby officials would have or find a graduate student or faculty member in the sciences, or I would have some one in place before Colby sells everything (in which event they cannot use the principal, only its interest for the continuation of the collection in my name). Also involved is my age of 84, coupled with a series of financial and court suits—including two women craftspeople owing me $4000, and a local wheeler-dealer ... who, in the course of falsely declaring himself bankrupt while being sued by four separate cases—plus one of mine—is also suing me with amounts of $225,000 and more involved. The whole, incidentally, is the deep-seated malaise of our times.

MA: *There is an enormous underground of people who reject the paradigms of science, who have created free energy devices, who know for a fact that some of the most archaeological facts about our past are covered, and that many of the masonic conspiracy are responsible for how this techno-civilization is withering away.*

BP: Each day that passes I am aware of native peoples to this area of the east coast (Maine, New Brunswick, Canada, Labrador, Greenland, North East Territories). It has been inhabited for fifteen-thousand years before Europeans arrived here in the year 1,000. 22 Salmond Street of Belfast Heights, as it is shown on early Belfast maps, is 171 feet above low mean tide—which 15,000 years ago was under water with the coast of Maine 150 miles to the north near Bangor. As the land rose, the waters receded, an entire plateau of living space developed on which people devised techniques of survival from, of, by nature itself. Rather than copy these obvious methods, arts, tools and procedures we have chosen to forget and ignore them in a mad, money-making pursuit of fouling our own nest with monstrous destroying gadgetry that will ultimately devour us. Our senses, teeth, bodies are defiled by toxins, radiations, residues, infestations in the air we breath, the food we eat. Behold modern man, king of his mess.

MA: *Has the Institute researched ethical systems? In my experience many young anarchist people are without any deep-seated belief system because they have categorically rejected all institutional belief systems. We at Dreamtime Village are exploring a plant-based ethical system based on the information inherent in plants as teachers.*

BP: For years as a group we have, we feel, explored all systems and continue to do so without yet feeling or sensing we have achieved a satisfactory result either for ourselves or others. Please continue earnestly this valid plant relation of which we whole-heartedly approve. We wish you measurable success. Plants preceded both man and animals by many, many years and in their growth and survival patterns lie many useful fundamentals, all adaptable to our daily use.

MA: *In genetic engineering they are taking genes from plants and inserting them in animals, etc. If everything is nature then this is yet another evolutionary twist. How do you see this problem?*

HOLD ON TO YOUR HAT

BP: All states or conditions or stages are in a condition of continuous evolutionary change, and man manipulated rearrangement in random, often uncontrolled means. As for the current moment, we must know and understand as completely and fully as possible all past, present, and future aspects. The how and why of evolutionary process must now be fathomed by whatever means.

EW: *Are there any women members of the Institute?*

BP: Only one, and all members are for life. Janelle Viglini, a San Francisco poet, heads the literary prose and poetry department. We know of no one who could replace her or double her department activities. Gertude Stein, Anaïs Nin, and Erica Jong[1] are honorary members. Over the years five to nine women on the average apply for Institute Membership, compared to one to four average for men. To both sexes we say neither yes nor no but rather: continue earnestly on the work you have outlined and submitted now and return in three years. Further facts reveal no one does. Obviously in this array are many, many worthy ideas, some of which we have either already worked on or are now engaged in pursuing. Frankly, there are more than many we would like to adopt as our own but so far have not succumbed for honor's sake, and feel if the applicants do not return to us in three years they may later, or we will hear of them through other sources. Finally, a very substantial number of women in pornography—movies, tapes, cassettes, photos, clothes, fan clubs, phone—come to us, not as joiners, but feeling advanced thinking produces ultimate truths we will share by mail.

EW: *Do you think there are fundamental differences between the male and female intellects? Would the world do better with women leaders, or best with no leaders at all?*

BP: In the Porter System of Simplistics
there is no Congress
no Senate
no legislation
no Congressmen

---

1. Erica Jong (b. 1942): American poet, novelist and essayist

no Senators
no Legislators
no lobbyists
no Democrats
no Republicans
no Independents
no caucuses
no other splinter parties
no chambers of deputies

Both nation and state-wide. The President and governors would be business-orientated women with four business-oriented women cabinet members in touch at all times with the people by modern technological communication systems. This system for the country and the states is based on the incredible and not fully understood or accepted difference in the male and female intellects with women the superior. For years and from the very beginning the Institute has advocated women leaders.

EW: *Do you believe in chaos? Is the universe an entropic or a non-entropic entity? Do any members of the Institute work with chaos theory?*

BP: Yes
We believe
in chaos
the random
the uncertain
the unsupposed
the unpredicted
the unknown
the yet to be discovered

The Institute sincerely endeavors to embrace all these categories in its studies and with the understanding we have yet a lot to learn and know as basic fact. Further, we are aware there are things we will never know, indeed are not permitted to know.

HOLD ON TO YOUR HAT

EW: *Have you or any members of the Institute communicated with plants? And if so, what have the plants told you?*

BP: Yes. They ask us
water us
protect us
let us live
save us
help us survive the terrible destructive forces and effects that technologies have forced upon us. The five acres of property here on Salmond Street abound in sparrows, robins, ravens, gulls, squirrels, raccoons, ground hogs, crows and others. We are aware that all of them communicate freely and daily with one another. We converse with each other and all of them. All of them say *Please save us from the horrors being inflicted on our air, on our water, on our food, on our bodies. Please save us—all of us.*

EW: *Do you or any members of the Institute have any involvement with aboriginal peoples? Do you think, as some do, that the Australian aborigines are the only true human beings on the planet, all the rest of us being only simulacra?*

BP: While working in Tasmania on methods to produce paper from hard wood and traveling widely in Australia I found the aboriginal peoples, with their techniques of survival, similar in every way to the techniques of survival employed by the natural five tribes of native peoples who have lived in my part of the world for 15,000 years. No difference. The major sorrow is that we bystanders have chosen to mock and ignore their ways. They could teach us much if we would listen. What matters is that original peoples are superior to us. USA—only 350 years old in the midst of cultures

Interviews with Bern Porter

6,000 to 15,000 years of age—have produced little beyond apple pie à la mode, a few medicines, and vehicles that fly in the air.

EW: *Why or how did human beings begin to go astray? Or were we a race always bent on destruction?*

BP: The entire full-length history of man reveals he has been ever curious, always adventuresome, eternally seeking, searching, finding. Even Adam and Eve in the Garden of Eden were fooling around with apples and snakes. But ever-sadly the end results were and are highly destructive in the rush to achieve without consideration of costs and after-effects. Or as we say aptly, man has fouled his own nest.

EW: *What steps will the human race take towards eventual non-use of language as we know it? Will we communicate in sounds? Telepathically? What can we do to prepare for this language-less future?*

BP: Merely realize the inevitable is in progress daily. That our five senses are slowly departing us. All we have left is plasma, sending and receiving impulse grains of meaning. Relax to it, prepare.

EW: *Do the members of the Institute, spread out as they are across the earth, communicate telepathically?*

BP: Yes. Knowing one another's plasma frequency of sending and receiving, each of us merely turns to one another's frequency just as you turn the dial of a radio or TV set to receive a certain known and wanted station or program of publicly known frequency.

Ben Meyers: *Have any of the ideas of the Institute been published in other scientific or socially-oriented publications?*

HOLD ON TO YOUR HAT

BP: Yes, but in an embarrassing, limited way. I say embarrassing because like [Wilhelm] Reich, we are always somehow ahead of our times, and editors find it easier to dismiss our submissions and us as crazy, stupid, incompetent crackpots.

BM: *Are there more simple or obvious ways to communicate ideas effectively than through direct statements? For example, meaning communicated diffusely, rhizomatically?*

BP: Yes, by the inter-frequency play already noted above. What can start as dots and dashes as in telegraph systems can reduce to continuous strips of communication energy.

BM: *In a society where everything seems bent on reducing perception to a single sense (if even that), does the active cultivation of all five senses constitute a subversive act? If so, does this kind of subversion become any more effective by being labeled as such, i.e., being made obvious?*

BP: All systems, no matter how well intended or designed, are rich in unnoticed subversive elements. Bear in mind the five senses are being biologically diminished.

    Hearing needs aids
    Seeing needs aids
    Smelling is mostly gone
    Tasting is disappearing
    Feeling is about through
    Food is contaminated. Air is fouled.

Subversion is highly rampant–effective. Note that for horse-drawn sled rides now offered in Maine, the trails, sleds and horses are contaminated by toxins, industrial wastes and radioactivity. One subversion reduces to two others. One fouled condition becomes more of the same, resistant to clean-up.

Photographs by Joel Lipman

HOLD ON TO YOUR HAT

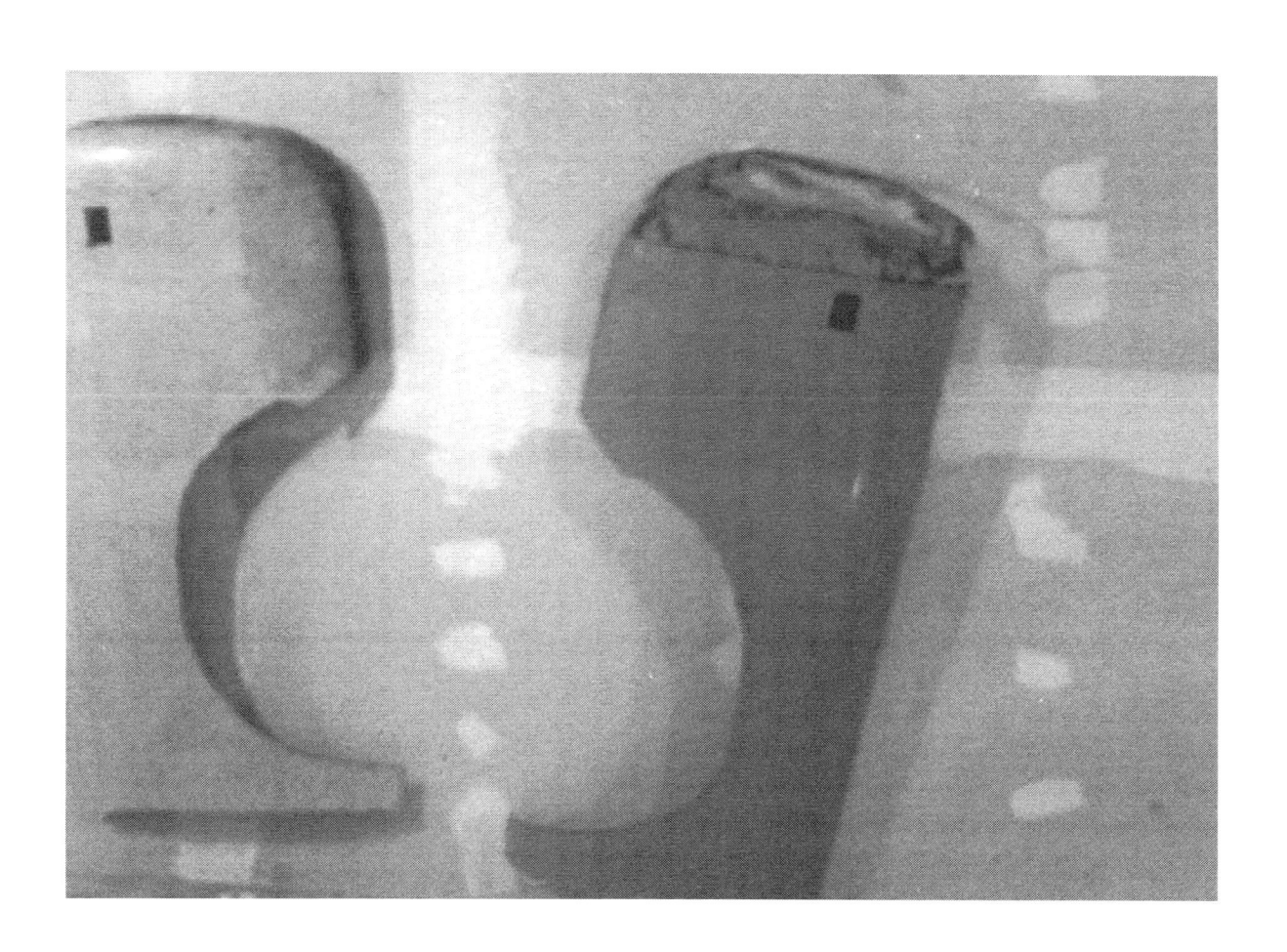

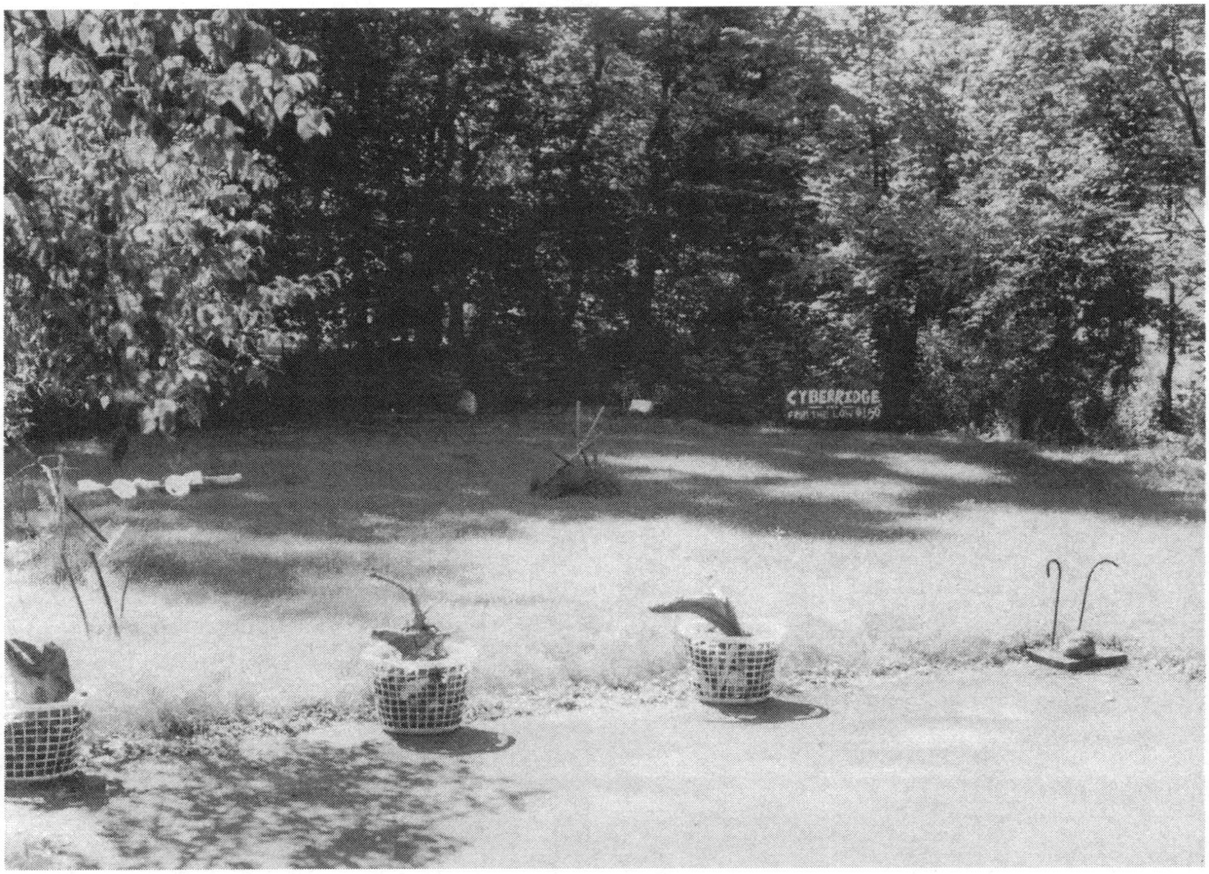
CYBERRIDGE

HOLD ON TO YOUR HAT

Interviews with Bern Porter

HOLD ON TO YOUR HAT

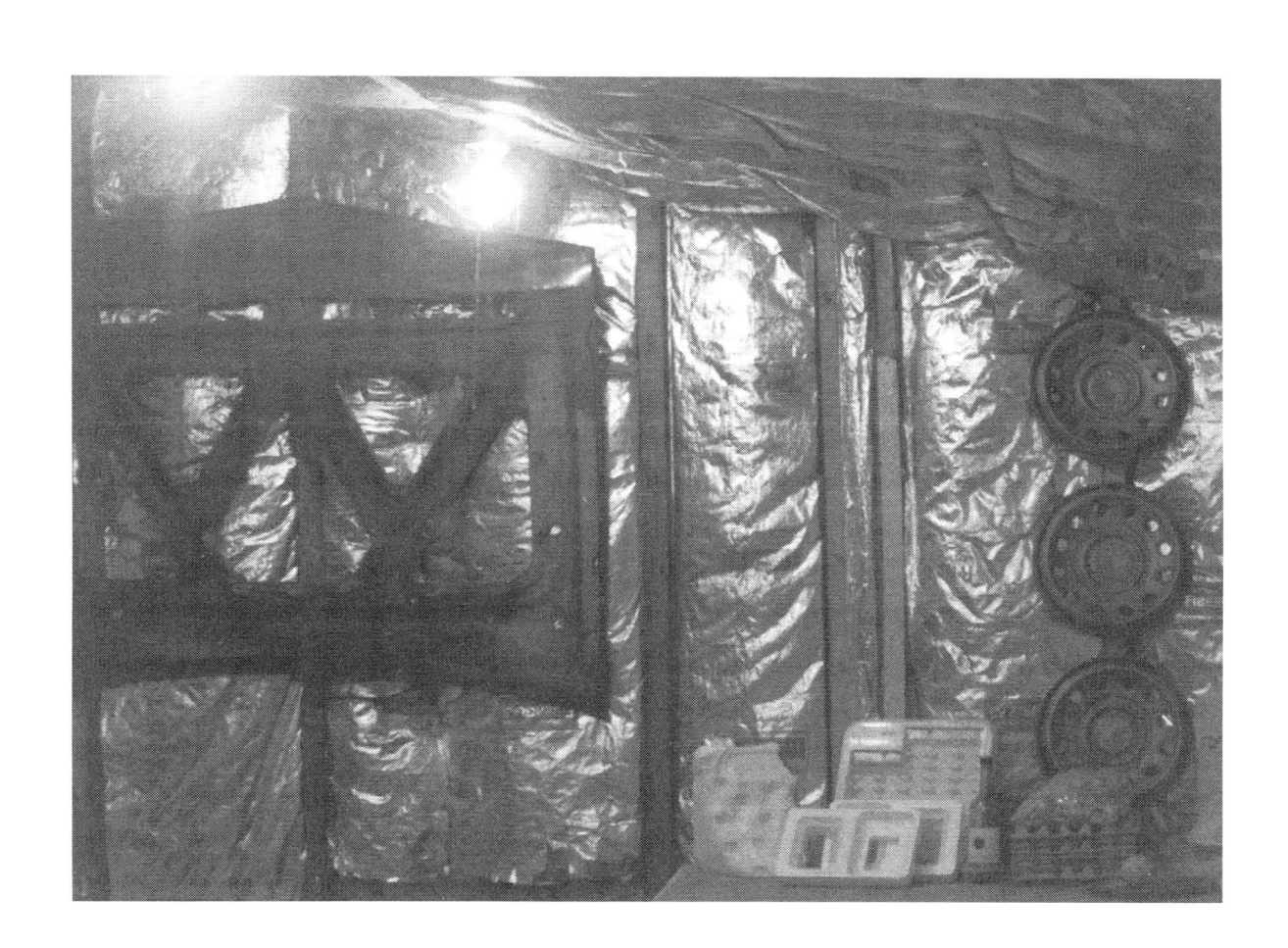

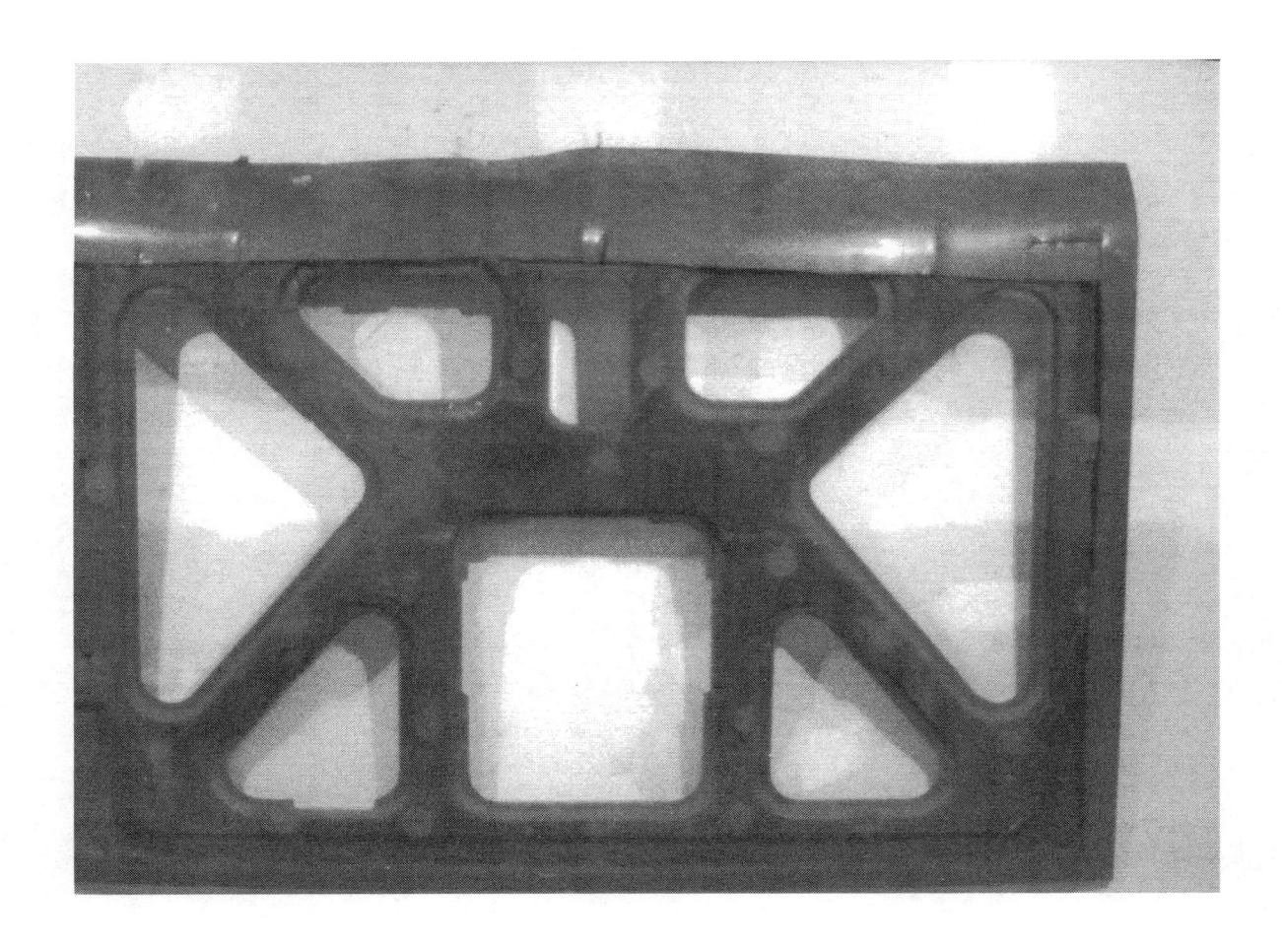

Interviews with Bern Porter

HOLD ON TO YOUR HAT

Bern Porter interviewed by Dick Higgins
Woodland Patterns, Milwaukee, Wisconsin
March 16, 1990

Bern Porter: We're in the exhibition hall of the famous bookstore. We're sitting around discussing which end is up, and if it's not up, then certainly it must be down. My friend Dick Higgins is just here and he will now speak.

Dick Higgins: *I'm Dick Higgins and I lived in Milwaukee for a few months once in '77. I've been back since and I've known Bern for thirty years. One thing I've never known about you though, is, how did you meet [André] Breton?*

BP: Well, it happened that he was in the basement, basement being the cellar, and he and what's-his-name had just had a very tremendous argument. They were trying to formulate a formula for the resurrection of life, how it could be restored and be brought back to this part of the world.

DH: *This was in Paris, right?*

BP: This was in Paris, Gertrude Stein was down the street a ways. She was running a Greyhound bus terminal for intellectuals. [Laughter] They were streaming in and streaming out, there was old Ernest Hemingway, there was so and so, and André was down in the basement three doors away, very, very angry and upset. He says, Gertrude just threw me out! So I said, well old man, forget about Gertrude. You and I will carry on. We will resurrect the world. And as far as I know we did that.

DH: *Mm-hm. That was about 1936?*

BP: It was, it was, and I remember Gertrude being very very very. She was always very as far as I could see. And André was . . .

DH: *That was before her trip back to the US?* [1]

BP: Yes, yes, she was always threatening to come back to the US. And no one ever could make up their mind including her. Did she hate the U.S.? *Do I hate the U.S.?* she used to say. But the U.S. intellectuals, so-called, were coming through the back door, they were crawling through the underwear....

DH: *She was pretty jealous at that point, wasn't she, because of all the success that Hemingway had had, and she felt kind of ignored until she made that American trip.*

BP: That's for sure, her feeling was, *Look, I'm supporting all these characters coming through, and they're not doing anything for me. Why don't they invite me back to my homeland?* So I said, *Well look, Henry Miller is over here, why don't you do something for him? Well,* she said, *Dear old Henry, what can ever be done for Henry?* So I said, *Well, you might throw him a couple of bucks, you might give him a meal ... How about a little half a bottle of beer? That would set up Henry very nicely.* She said, *Oh no, we can't, we can't subsidize Henry. We can't subsidize Henry that way.*

DH: *Did you see Breton when he was in New York during World War II?*

BP: Well, he came, there was Peggy Guggenheim here you know, she brought the boys over. She took off from Portugal with a planeload of them. She brought them

---

1. Gertrude Stein's American Tour took place from 1934-35.

all to America to escape Hitler. And she set up her place, Art of this Century, a very extraordinary exhibition/meeting place. She had salons at her apartment. She really saved the world there for a few months. Brought the intellectuals together. And among them of course was our friend André.

DH: *I went to school with his daughter in Vermont, so I was fairly early, all things considered in my life, to be aware of Breton's activities.*

BP: Well, he has set a stage, I think, of influence for all the rest and it's too bad that now with the passage of time, few if any remember him. I'm grateful for your mentioning him at this point.

DH: *Oh, I thought it was good to get some memories of that. But weren't you based in California at that time, or were you somewhere else, that would be just before the Manhattan Project.*

BP: Well, I was in New York City at the time. I was a physicist for the Atcheson Colloids Corporation. I was at 444 Madison Avenue. And nights I used to go down to an art school set up entirely for Italians. I used to go up to see Hilla Rebay[2], I used to see the Guggenheim collection, and I used to see Peggy off and on. In fact, I took one or two of my constructions from the physics laboratory at Princeton, and I said, *Look Peggy, I'll either give this to you or you can pay me 20,000 bucks for it. Which would you prefer?* [Dick laughing] She looked at it for a while and looked at me and looked at it again, and said *Well, Doctor—Doctor Porter*, she called me—*I'll consider it, I'll consider*

---

2. Hilla Rebay (1890-1967): French-born painter and first director/curator of Guggenheim's Museum of Non-Objective Painting

HOLD ON TO YOUR HAT

it. Now so there's no misunderstanding, she said, *I have two choices: you either give it to me free, or I pay you $20,000 for it.*

And I said, *Yes, Peggy, that's the deal.* Well, I never heard from her since.

DH: *[laughing] How long did you stay in New York?*

BP: Well, I was there for five years.

DH: *Roughly '41 to '46 or something?*

BP: Something of that sort. I'm all hazy at this time about dates; maybe it was more like '36 to '40. In any case, those were very rich years. There was Mabel Dodge Luhan³, came in from Arizona, she and her Indian boyfriend [Tony Luhan]. They wanted to set up a salon and compete with Peggy. Peggy was fighting with someone else. There were these intrigues and underground movements. Clashes of personalities. Who was bigger than whom.

DH: *[laughter] Were you close at all with the circle around* View *magazine?*

BP: Well, that was a very rich period. There was Charles Henri Ford⁴. And *View* magazine according to me, was not only a kind of pioneer, it was a pioneer. It set the stage for all since. There was Parker Tyler⁵ and there was Charles himself, and they set up a museum, they set up a magazine, they took part in the 1937 World's Fair. I've forgotten who—I think it was John Cage—who supplied 'em the money to set up this exhibit.

DH: *No, he was poor as a church mouse at that point, it can't have been him.*
BP: Well, there was someone, I forget the name. All I do know is at this point that

---

3. Mabel Dodge Luhan (1879-1962): American writer and salon hostess
4. Charles Henri Ford (1913-2002): American poet, novelist, photographer and editor of the magazines *Blues: A Magazine of New Rhythms* and *View*
5. Harrison Parker Tyler (1904-1974): American poet, novelist and assistant editor of *View* magazine

Charles and Parker Tyler set up a pioneering effort that, as far as I'm concerned, is still going on. I subsequently published Tyler's *The Granite Butterfly*. And Charles inserted ads of mine in his magazine and all these ads had to do with what I call Metamorphic Rhizomes. That is to say, decaying roots. These I mounted on plastic bases. I collected them while I was a member of the Intercollegiate Alumni, a weekend group that went up to Bear Mountain, where we cavorted through the mountains and entertained ourselves with exotic stories.

DH: *[laughter] Remember any of them?*

BP: Well, yes. The most exotic of the stories was about a wounded bear. It seemed a wounded bear climbed a hollow tree and when he got to the top, he was encountered by a nest of bees, a beehive, in other words, a natural beehive. And the bees came down and they intrigued and fought among themselves as to who would inject the first bite on these intruders. There was a war between the bees and the men. An exotic situation which I often think of because I think we're still at war, the bees against the men.

DH: *Mm-hm. And then when you went to work on the Manhattan Project, did you remain in New York?*

BP: Well, in those days we were shipped off first to the University of California at Berkeley and we were shipped off to Oak Ridge, Tennessee. I use the word "shipped off," and I think that's it. We were treated as pieces of baggage really. The interesting thing was that on the first trip between Berkeley and Oak Ridge, I stopped off in Chicago. I went into Ben Abramson's bookshop, where Ben was, under the counter, selling copies of *Tropic of Cancer*. He was bringing them in from Paris. And when I showed up at the front door and said that I was interested in *Tropic of Cancer*, he said to himself, I could almost hear it, *Here's another spy, here's another intruder, he's gonna turn me in, he's from the immigration department, from the censorship department, this guy.* Whereupon I said, *No, none of that is true. I have a letter from Henry. Henry himself has written this. Saying that I have his permission to come here* *to your store, and he has given you permission to sell me a copy of* Tropic of Cancer—this was an underground operation, whereupon Ben said, *Well, how do I know that this note signed by Henry is not counterfeit? How do I know it's original?* I said, *You can call him up, he's living in a garage, just go in and ask him if it's ok.* He said, *Well I can't do that, but,* he says, *you look like you have an honest face. And besides that, there's something about you that gives me the impression you're a literary guy. So maybe I'll take your word for it, and maybe if you give me some bucks, I will exchange the bucks for a copy of* Tropic of Cancer, *provided you do not tell anyone under any condition whatsoever, where you got it, how much you paid for it, and who sold it to you. This is a secret confidential underground matter; the sale of a* Tropic of Cancer *in this backward country is forbidden by law! If you're caught bringing it in the baggage you can be arrested.* I was on the point of saying, *Well look, Ben, how did you get them?* But I didn't. So I gave him three bucks and he gave me a copy and I raised my right hand and I said, *Ben, thank you. I will never ever tell anybody where I bought this.* He said *Bern, please get out, I'm busy.*

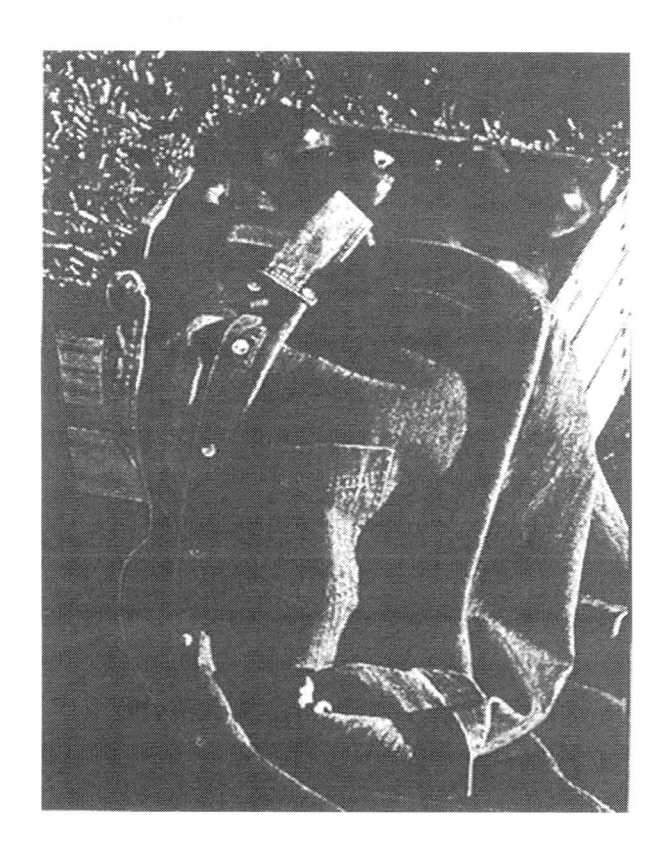

DH: *[laughter]* That's a great one, Bern. So you had a copy of Tropic of Cancer *with you when you were working in Oak Ridge, Tennessee?*

BP: Yes, yes. And out through the back door, I would go in through Oak Ridge or the next town and I would arrange for the printing of some of Henry's works. The first thing that we printed there, at Oak Ridge, as I remember, is *What Are You Going to Do About Alf?* Having printed it, I discovered, or someone called my attention to the fact, that in this little volume, two and a half inches by three inches, there were some four-letter words. Not only were there some four-letter words, there were quite a few. [Dick laughing] And so I kept thinking about this and thinking about this, and I was thinking also about what Ben Abramson had said, how all this stuff was underground, it was censored, and under no condition should anybody see it or know about it. So when I went home to Maine on a short vacation, I took copies with me, and I went out to my grandmother's cottage on Nickerson Lake outside of Houlton. With India ink I personally blanked out and covered over a staggering number of four-letter words.

DH: *[laughter] That's funny.*

BP: My grandmother asked me, *Well, Bern, boy, what are you doing? You seem to be very busy!* I said, *Yes, Grandma, I am guilty of publishing a book that has four-letter words and now my conscience bothers me so I am covering them over.* And the sequel to this story is as follows. Eleven copies were made, covered over by ink. And I've been told that if one of those eleven copies can be found, there are people who'd pay twelve hundred bucks for one of them. [Dick laughing] So my problem is where are the eleven copies? And where, how, what could I do if I had twelve hundred bucks?

DH: *Well, you might look at Houlton, Maine!*

[laughter]

BP: Ah, dear old Houlton.

DH: *Yeah …*

BP: Well I was born out in Porter Settlement, which is about three miles east of Houlton. My mother wanted to come into the big town, the big town of Houlton. She didn't want to be associated with farming, so we moved into Houlton, where I began my education. And I would like at this point to establish that before I moved into Houlton with my parents, and before I left Porter Settlement, I single-handedly with my right hand, with my own pen and with my own ink, established and set up what is now the phenomenon called Mail Art. I claim to be the inventor, the originator, of Mail Art. Then about three years later, a guy named Marcel Duchamp in France says, *This guy Porter in Porter Settlement, Maine is a fraud. I, Duchamp, invented Mail Art.* And to this day there's this historical confrontation. Was it Duchamp or was it Porter? Duchamp says, It is I. Porter says, It is I, and I'm not sure that the problem will ever be resolved. And I repeat as I sit here, at the Woodland Patterns in Milwaukee, in the year 1990, that it was Porter, Porter that invented Mail Art.

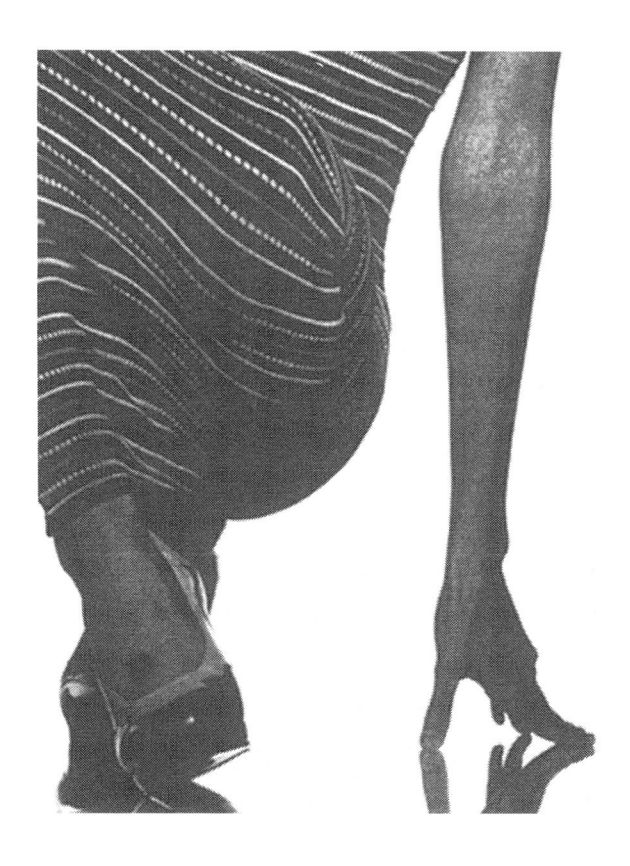

DH: *Mm-hm. Who called you "Bugs"?*

BP: Well, in those days, everybody had to have a nickname. Even today I understand that it's not acceptable to be calling people by their name, you have to invent it. And among teenagers this is a hobby.

DH: *And it was a hobby then, too! [laughter]*

BP: Yes, it was, it's a deeply entrenched hobby, a deeply entrenched hobby. Everyone should go through life with a nickname. Well, so. Here comes this guy Porter. He was wearing golf knickers, [Dick laughing] caddying summers at the golf course, and he used to show up at school in the fall wearing golf knickers. So it was very clear that this guy was in golf clothes. And the long-handled underwear was really far out. And the problem was, *What are we gonna call him?* Well, it was quite obvious, just call him "Bug." B-u-g! What a name, Bug Porter.

DH: *But it didn't last very long, did it?*

BP: Yes it did, it lasted for, well, I was named I think around 1917 or so, and I must have carried it through to 1924. And it's interesting that when I got to graduate school to study physics at Brown, there was another man whose first name was Bernhard, and my name is Bernard without the h. So the problem was how do we distinguish between these two characters? So they solved this tremendous problem in a very logical fashion. They called him Ben, B-e-n, and called me Benjy. [Dick laughing] So there we were, Ben and Benjy, setting out to transform the laws of the world's true physics. So in my Bug period I had already invented Mail Art, and in my physics period as Benjy, I'd set up the makings of the destruction of the world.

DH: *When did the saxophone come in?*

BP: Well, the saxophone for me was really a clarinet. My parents were able to get for me a clarinet. And I was able to study with a cousin of mine, he and I would go to this man's house at night, sit in his kitchen and learn to play the clarinet. Having learned, we then became members of the Houlton band. Then I went on to Colby College where I continued in the band, and kept the clarinet for many years. As a matter of fact just this past winter or so I finally gave it to Democrats in Belfast for an auction set up to raise money to support the Democratic party.

DH: *Aw, Bern, I wish you'd sold it to some collector ... [laughing]*

BP: So the Democrats sold it at auction. I had it for about fifty years. And I must say I was a virtuoso on the clarinet, I must say so, in all candor.

HOLD ON TO YOUR HAT

DH: *Did you compose any pieces for clarinet?*

BP: Well, I tried to, I tried everything trite in my time. And it's possible that some of the greatest scores ever written for the clarinet are now in Westwood at UCLA, whatever it is.

DH: *In the library that has your artworks and early works and things.*

BP: Meanwhile, I think you and I should talk about our association.

DH: *Oh, we'll get to that. But I'm just enjoying some of these miscellaneous things, like Niels Bohr, you worked with him too.*

BP: Well, good old Niels, he was very envious of us Americans. If he came to Berkeley at lunchtime at the faculty club, I remember he used to say, *Why is it that you Americans don't recognize Swedish science? I'm the epitome of Swedish science.*

DH: *Even though he was Danish?*

BP: Well, there was a confusion. Was he a Swede or was he a Dane?

DH: *He's Danish.*

BP: I asked him that one day. *Well, he said, we're so close together kin-wise, blood-wise,* [Dick laughing] *how can you tell the difference? We all slept together in the early days....*

DH: *It's funny, because the Danes and Swedes were traditional enemies.* [laughter]

BP: Yes, but in the field of physics, they agreed that they would put aside all their angers, their inhibitions, and they would really be pioneers. It turned out Niels made quite a contribution, he studied the molecules, the structure of certain elements, examined many things, went over to Paris and saw Madame Curie. Curie, her husband

and their in-laws, they were fooling around with pitchblende, cooking and stewing and setting up what is now called the radioactive theories.

DH: *But your specialty was more like statistics or something.*

BP: Well, my specialty was trying to keep everybody on an even keel, [Dick laughing] and trying to get them all to agree, is there something more we can do with this stuff than what we're doing? For example down at Oak Ridge, weekends we would go off in the woods and set up a bonfire, cook lunches over the bonfire and sit around and pound on our chest saying *Well, we're really great guys. We're making atomic power for trains, for submarines, for airplanes, mobile plants. We're making contributions to agriculture. We're promoting the growth of botanical specimens, we're finding tracers for the human body, we're running in the forefront of the cutting edge of a new technology. We're really bright stuff.*

DH: *But you quit that project around '44 didn't you?*

BP: Well, as it turned out, there were something like seven methods for separating uranium. One was used, features of the other six were not really explored. And the forty-two uses that we cooked up were set aside and some people up in Washington sat around and said, *Look, we have to go over and put those guys out of business, and we have to do it with one blow, and we have to be final and make a final statement. How shall we do it? Well,* said one, *we could just take off from Saipan, an island in the Pacific, and go straight ahead and drop it at the first point.* Whereupon someone said, *I don't think you should take that path because on that path are some of the treasures of Japan. Could you please straighten the line of attack a couple of degrees to the south? And then having spared the treasures of Japan, its temples and whatnot, just go straight ahead from Saipan at a certain speed at a certain height and when the clouds open below, drop it.* So they took off from Saipan. And the story goes that the first place they came upon was Shimonoseki, which was known to be the home of Christians. Baptist missionaries had come there in the early days and had persuaded all the members of this town who were involved in production of textiles to be Baptists, to give up their

HOLD ON TO YOUR HAT

Confucius and whatever and whatever. The story goes that the people in the airplane coming over were also Baptists, and the intervening layer of clouds separated the two Baptists, those on the land and those in the plane. The plane continued on its journey at a certain level and a certain speed, and the open space was Hiroshima. And they dropped it.

DH: *And you went on ...*

BP: I came later. And I would sit at the bars in Hiroshima. The guys would say to me, *Why did you pick on us? Why'd you drop it on us, we didn't do anything! We're nice guys. Furthermore, here in Hiroshima we didn't have any ammunition plants, we were not making bombs.* The only thing I could say was, *Yeah, but you guys came over, you hit us in Hawaii. And you turned on a war against us. Yes*, the guy said, *we did that, we did all that. Sure, we're guilty, but the Emperor told us to do it! And when he speaks we do what he says.* And then one smart guy over in the corner of the bar, he says to me, *Instead of dropping it on us, why didn't you drop some leaflets saying that tomorrow noon you would make an island five miles off of Hiroshima disappear before our eyes, as a demonstration of what you have, what you can do with this thing you call the bomb, b-o-m-b? Well*, I said, *yes, that's an idea but we didn't do that.* Whereupon he said, *Our Emperor is an internationally-known botanist. For years he's had a hobby of botany. He's versed in scientific methods. Why didn't Einstein come over and sit down with the Emperor and with a piece of paper explain to him what was in this bomb. Why*

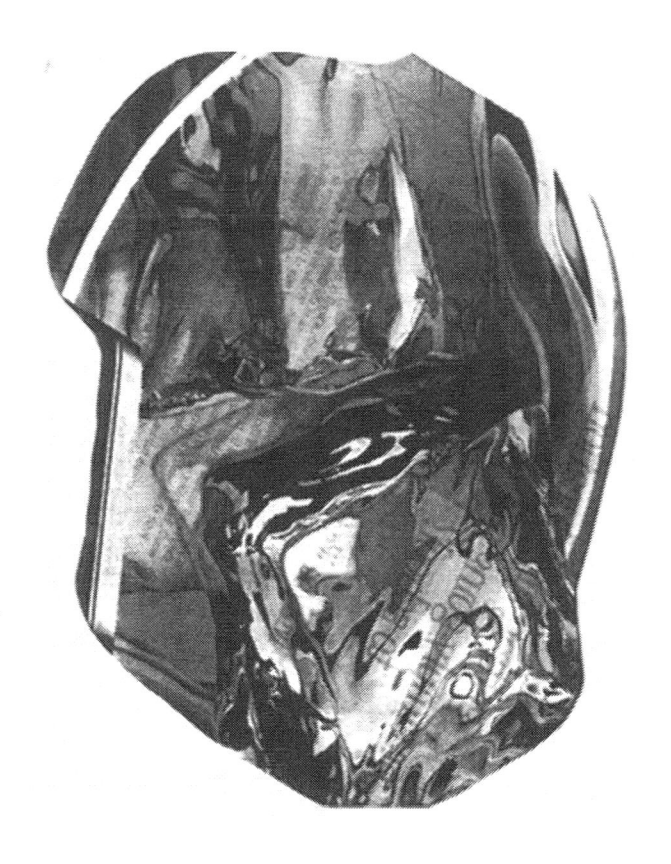

Interviews with Bern Porter

didn't you give us a warning, a demonstration? Tell us, why did you hit us? We're a peaceful people. Well, yes, I said, you're a peaceful people, but you got a rationing system with sugar and flour, your mothers are sending you guys off to war, and you're taking part, you're guilty. Oh, well, yeah, but the Emperor told us to. [Dick laughing] *So, so, anyway you dropped the bomb and here we are and what are you doing here now? Well,* said I, *my conscience bothers me and furthermore I'm a part of this crowd of Americans who have come here and have said, "look, we bombed you, could we now please examine you? Could you please tell us what we did to you? Would your women come in and be examined, could we, and you boys, would you please come in and submit to our questionnaires? Where were you when the bomb went off? Were you under a roof? Were you lying down? Were you sitting in a chair? Were you near a wall? What clothes were you wearing? What's your diet? What's your hereditary background? Let us examine you."* Well, the women took part but the men didn't. And then of course five days later, they came over to Nagasaki and dropped another one. And historically the question is, having dropped one, why should we drop two? Another historical question is, since they bombed Hawaii, why don't we just let 'em have it? Why do we go to war? If they want Hawaii, let 'em have it. We don't have to fight. Well, other interesting things, we're at Woodland Patterns now in Milwaukee. [Dick and Bern laughing] The people of Europe have discovered that, after seventy years, Communism doesn't work. They want peace and democracy.

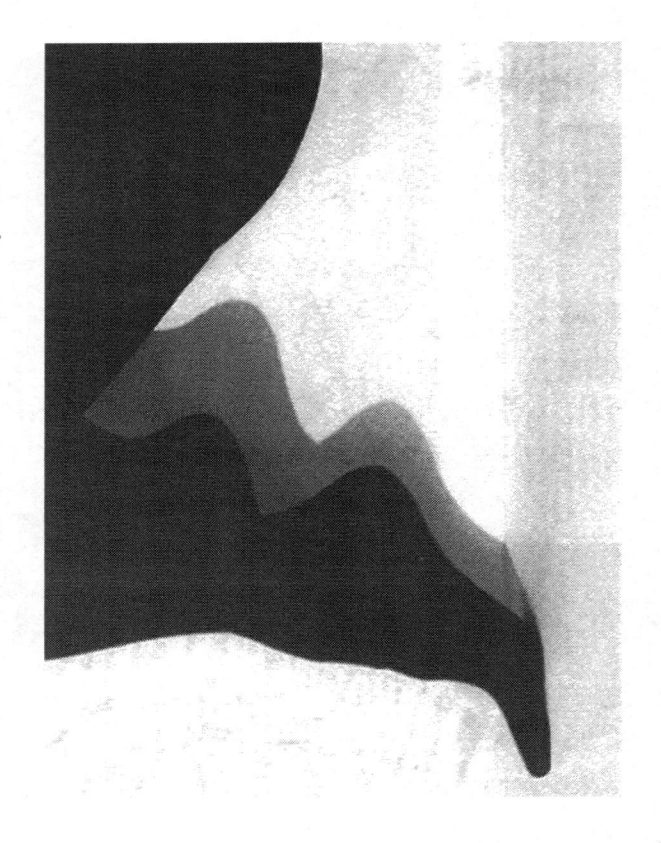

DH: *Well, so we hope.*

HOLD ON TO YOUR HAT

BP: And here we are, we would like to have a little democracy.

DH: *So then, there followed your time of traveling around a lot. You went to Sausalito?*

BP: Well, this past winter, only two months ago, I was in Brunei, the smallest and the richest country in the world. Where the per capita income is fifteen thousand bucks. Where the Sultan in his generosity has given to every inhabitant in Brunei an automobile. He has given them a color TV. He has given them free education, free medicine and free food. The great Sultan of Brunei is taking care of his people. And interestingly enough, a large portion of them say, *Yes Sultan, thank you for the car, thank you for the stuff and the stuff, but we would just like to be living in our own way, in our own little shacks on piles over the water.* [Dick laughing] *But since you've given us the cars and stuff, some of us are sort of smart, we've taken our fifteen thousand bucks and we have created architecture the likes of which are nowhere else in the world, some of the most beautiful forms–spectacular units–for dwelling you could ever have seen or could ever imagine.* So, yes I travel, yes it's true.

DH: *Then you moved from Oak Ridge to Sausalito and from there to Samoa. You like the Pacific countries don't you?*

BP: Yes, yes, they have a personality of their own. It's sad they're disappearing. It's been called for years the paradise of the world. You see the pictures, you see the gals dancing with their beautiful hands. With that slowly, slowly going . . .

DH: *Were you publishing when you were in Samoa?*

BP: I carried on publishing activity from Kual, one of the islands of the Pacific.

DH: *And Tasmania?*

BP: Yes, Tasmania, dear old Tasmania. So here we are.

Interviews with Bern Porter

DH: *And then I made contact with you after your return to the U.S. when you settled in Calais, Maine.*

BP: Right, and we turned out a very fantastic item, what was the title by the way?

DH: *It was called* What Are Legends?

BP: That is correct, I remember now, I remember now. I illustrated . . .

DH: *Well, I'll tell you a little about that, if you don't remember.*

BP: Ok, yes.

DH: *Ray Johnson[6] was in contact with you through Mail Art.*

BP: Right.

DH: *And he was sending collages around and you were sending them back slightly modified to him, and he was very impressed. And you enclosed with one of your mailings to him a little brochure that you had produced back in Sausalito that gave a sort of summing-up of your career up till then. It would have been produced around 1947 maybe, and this was about 1959 and I was about to go to printing school. So I was delighted by the pamphlet that Ray gave me and I wanted to make contact with you, so I sent you some of my manuscripts. And you wrote back almost by return mail, saying that Dick Higgins being of sound mind and stout body, or something like that, would be welcome to contribute to Bern Porter Editions. Nobody had ever said that to me before, so I sent you another manuscript which was* What Are Legends?—*which is the theory essay that goes along with my Legends series which was much later published as* Legends and Fishnets. *And that one you then hand-lettered because we didn't have any money for typesetting and I took it to the printing school where we had it printed with your beautiful illustrations.*

---

6. Ray Johnson (1927-1995): Collagist, co-father of Mail Art and founder of the New York Correspondence School

HOLD ON TO YOUR HAT

And remember we had a big fight with various binders who said it was too small a book to have a square binding, but we finally found someone who would put a square binding on it, a square-back binding, because you wanted to make sure we could see the title. And it really looked absolutely beautiful, it's a delightful little book. I'm very proud of that first book of mine, which was also our first collaboration. And that came out in 1960.

BP: Well, so there it is, thirty years later.

DH: *I'll mention one thing more than that. Somebody told me that you had to have a picture of the author and the artist on the back cover of a book, and I didn't really know what to do about that. I asked you several times in letters for a picture of you, not knowing that you were not fond of having your picture taken. I knew you lived in Calais, and Calais is in Washington County. So I made a picture of myself and beside me I put a picture of George Washington, Bern Porter of Washington County after all. I sewed real buttons onto the picture and gave it to the camera people to photograph, which I guess is not on the cover but inside the back cover, and that's how that came to happen. Then one day the phone rang, and it was you, Bern. You said you'd just come to New York on a cheese box, you'd floated into the harbor, and I said well, anyone who talks to me like that I've got to meet. And I wanted to meet you anyway. So, you arrived and you did not appear to be very wet, you had not come into town on a cheese box. I had no idea how you did come. You were looking rather dapper in an old but interesting suit and nearby there was an Italian street fair. I asked you if you wanted to go to the street fair and you said no, no, I'd rather talk with ya. So, we talked instead, and I thought that was also very interesting,*

that he knew exactly what he wanted, this Bern Porter. There was no question what he wanted. You were here to meet me and talk with me and size me up and figure out what to do. The Italian street fair would have been a complete distraction. I could take other people to the Italian street fair but not Bern. So that was our first meeting.

BP: Very good, very good. I remember coming down at some point. Alison [Knowles] was there and I saw the two of you together. And for all these years, I have never convinced Alison to give me a manuscript.

DH: *Well, she doesn't produce very much writing. A modest amount.*

BP: Well, she does some fantastic pictorials.

DH: *Oh yes, indeed, indeed.*

BP: I've gotten a title. I have one of them, I'm sure. Well, at any rate, the years have gone by and ...

DH: *You went to Warwick, New Jersey and you went down to Alabama for a while.*

BP: Yes, I seem to have been gone, going ...

DH: *And then you came back to Maine. By then I was living in Vermont and it was very hard for me—since I was responsible for the children—to get away from Vermont, but I did manage finally to get them over to visit you in Rockland—you and Margaret, your wife— and that was an interesting thing. I remember eating out with you on a picnic table and I had been trying to get a picture of you. I was always trying to get a picture of you, but I couldn't get the right picture. You were tense and nervous or uncomfortable, it wasn't that you would say no, but you just didn't like it. So we got to talking and the camera was on a picnic table out at a beautiful park, and there were some nice rocks, you remember that park near Rockland?*

HOLD ON TO YOUR HAT

BP: Oh yeah ...

DH: *Oh, it was wonderful.*

BP: It's all we had, those wonderful rocks ...

DH: *Right by the sea, and we were having a great talk, and ...*

BP: Great talk by the sea.

DH: *And along comes my daughter Jess who was nine years old and she picks up the camera without either of us noticing and you were pointing out a rock out in the water that the waves were breaking over. And she ...*

BP: Yes, yes. See the rock! See the rock! I said.

DH: *That's right, and just at that moment she clicked the shutter and got the most beautiful picture of you I've ever seen, and it appears on the cover of this book here, I've Left. There you are, pointing out that way. It was really kind of a bluff so you were pointing down. And it's credited properly to Jessica Higgins somewhere here, but what it doesn't say is that Jessie was only nine years old! [Dick laughing]*

BP: Well, the years go by. I couldn't bear them so we published *I've Left*, which I considered my autobiography.

DH: *And you came to visit in Vermont shortly thereafter when we planned out this edition. Or maybe before you were with Margaret and she showed us how to make what we call, in my family, Margaret Porter onions. And that's a very simple recipe—you just take an onion and cut the top and the bottom off and put it in the oven on a cookie sheet or something like that (or you can put it around the roast) and cook it for about an hour and a half at no hotter than 350 degrees. And when the insides begin to pop out then you've got the most deliciously oniony onion you can imagine 'cause it's cooked right in the skin. And I'm sure it's very nutritious.*

BP: Yes, yes, well at Salmon Street or in Belfast I know I have a few of her paintings, her collages, her productions, and they're very much a part of not only my life, but of those who come there to visit. They spot these very unusual creations which she has left to me.

DH: *And so we published with Something Else Press two books of yours, Bern.*

BP: Yes, for which I am very grateful.

DH: *I've Left and also Found Poems. On that there's two photographs that I managed to get of you. We went to go up a small mountain and your shoes were very slippery and those were all the shoes you had with you. It had snowed a little bit the night before and the ground was very slippery. So you tried to go up the slope, you remember that day? Every time you tried to go up it was like a person beginning to ski, you would sort of slip right back and bump into a tree, and I took two great pictures of you that day. [laughing] One of the pictures I have over on my bathroom door. It's a huge thing that I blew up for a convention we were attending, and the other, the original at that, I just came across a few days ago in my files. And you looked sort of like Charles Laughton in Hobson's Choice, having a kind of disagreement with a lamppost. But it wasn't a lamppost and Bern you were perfectly sober, it was just awful weather and there you were with that big tree and laughing all the while. [laughing]*

BP: Yes, yes, well there were nothing but trees, nothing but laughs. And here we are at Woodland....

DH: *Oh and some other things that we talked about too. You told me about your Reichian[7] interests in that trip a lot.*

7. Wilhelm Reich (1897-1957): Austrian-born physicist who developed the theory of orgone energy

HOLD ON TO YOUR HAT

BP: So here we are at Woodland Patterns.

DH: *Yeah.*

BP: And on the wall we have some Do's and some Don'ts.

DH: *What's the difference between a Do and a Don't, Bern?*

BP: Oh, there is no difference, they're one and the same. Over here on the left it says, *Try it, it won't bite.* And over on the right it says, *Get a kick out of it.*

DH: *Never lucky to find in Indiana.*

BP: That's for sure. [Bern and Dick laughing]

DH: *Poor Indiana. And, don't look down on short men.*

BP: Nooo!

DH: *Nooo. There's no short men.*

BP: And over here, and over here it says, *Think negative!*

DH: *Yeah, yeah.*

BP: And another one suggests, *Go forward with!*

DH: *Yeah.*

BP: Listen to this page! *Think rats. Clean Sweep!* And over here it says, *Don't be so, don't be so, don't! Don't! Don't be so! You won't be.* [Bern and Dick laughing]

Interviews with Bern Porter

DH: *And one of my favorites in the* Found Poems *books is somehow in the same spirit. You have gotten a hold of an arrest list from a police station of what you can arrest people for, not the complete crime list, but the smaller things. And you took this list and made a poem out of it by claiming that you had done every single one of the offenses on the list.*

BP: And furthermore had enjoyed every single one of them and further would do it again if he had a chance! What a guy, what a guy.

DH: *That is one of my real favorites.*

BP: So, here we are.

DH: *Why do you use medical imagery so much in your things, Bern?*

BP: Well, the truth is that in the wastebasket in Belfast and in Rockland, all the doctors throw out their medical literature. They refuse to take it home or to their office or whatever. They just throw it in the wastebasket and I come along and I grab it.

DH: *Well, you could go by the butcher's, but maybe you wouldn't get such interesting images?*

BP: No, no, the butcher doesn't throw his things in the wastebasket at the public post office. Only material from the public post office goes into my books. I borrow my scissors, I make glue out of flour and water. You other artists you have easels and paints and canvases and presses and whatever and whatever and all I have is a wastebasket at the post office.

DH: *Well, here's the piece: "The 89 Offenses I've Committed, the 502 Times I've Committed Them, Plus the Implication That I Enjoyed Myself Thoroughly With Every Act on Every Occasion and Would Repeat Any or All of Them Whenever the Chance Arises."*

HOLD ON TO YOUR HAT

BP: So here we have aggravated assault, we have the attempt to illuminate wild game, and we have attempted larceny, attempted larceny of a moving vehicle, assault on a police officer, assault with intent to kill, assault and battery, break and entering, break and entering in the nighttime, breaking and entering and larceny in the nighttime, we have bench warrants, break and entering with intent, break and entering and larceny, behaving in an incorrigible manner, well I did that once [Dick laughing], break and entering in the daytime, concealing of stolen property, conspiracy, cheating by false pretenses ... [tape cuts off]

DH: *And it lists all the times that that was done within whatever period it originally came from in, presumably, Rockland, Maine.*

BP: Right.

DH: *Well, you moved up to Belfast to be closer to your brother, didn't you?*

BP: The truth is, I was traveling about the world and my brother was an optometrist there in Belfast and agreed to store my worldly goods while I traveled. He further agreed that as my mail accumulated there he would send it to me, to Japan, to Australia, to Norway, to Russia, wherever I might be.

DH: *You were the ultimate publisher on the move. [laughing]*

BP: Yes, well, my brother's now retired and still in Belfast.

DH: *So you do have one relative nearby, that's great.*

BP: Yes, it's very interesting. In a town of six thousand people, we move in two totally different circles, circuits. He's in with the Rotary Club, the curling club, the church club, the bank club, the whatever whatever. And I'm off to the side, I'm contaminated by poets and writers and dramatists and culturally oriented persons, who come by with sleeping bags and ...

Interviews with Bern Porter

DH: *But you still have your scientific contacts too, Bern.*

BP: Yes, yes, they also come and we have the Institute [of Advanced Thinking]. The Institute concerns itself with world problems from the point of view of physics. Now that Bern Porter Books has been shifted from Maine to California I have full time now to devote to the scholars and principles which we helped carry on all these years. What principles? Well, it looks like physics can do other things than destroy. Physics has use for the humanities, for music, for art, for communication, for sculpture, for religion. It's basic to all endeavors. It has nothing whatever to do with the horrors of destruction. Only the politicians diverted it that way to our sorrow. So we carry on, we carry on.

DH: *For example, you were down in Massachusetts to see [Ernst] Dombrovsky[8] and to go to a conference a few years ago. You came back and reported to me on how people didn't seem very interested in the ideas of physics, they were only interested in the qualifications of the people attending the conference. And every time you would ask to say something, they would say, And what university do you teach at, Mr. Porter? And you didn't like that too much.*

BP: No, because I could not answer them. I have never taught at any university. I am allergic to universities and they are allergic to me. To be a scholar at the Institute you have to prove that you are not affiliated with some university, that you are a free spirit, a freelancer.

DH: *You're of the same training that people of the universities have, even better, but teaching ...*

BP: Well, yes, but I think, I think they really ruined me. I should never have gone.

DH: *And you also have a Reichian interest too. Still?*

---

8. Leonid A. Dombrovsky (birth unknown): Russian physicist (affiliated with the Institute for High Temperatures of the Russian Academy of Sciences, Moscow) and photographer

HOLD ON TO YOUR HAT

BP: I still am involved in this. I constructed in the backyard of the Institute in the woods area, as far as I know, the only horizontal orgone accumulator in the world. Most of the ones that I'd ever seen consisted of a blanket or a telephone booth-like structure or a box. But this is a horizontal plane measuring something like twelve by eighteen feet. It has five sides, each side constructed according to Reich's principle of alternating layers of metal and wood. And from what would be the sixth side is the earth on which a record of some ninety years shows that a woman had lived there alone in a cabin of those dimensions, twelve by eighteen. And then when she left it became a rabbit hut and later a chicken and rooster place. In other words, this spot of twelve by eighteen feet has what Reich called the energy of habitation. The Indians are very much concerned with the energy of habitation. Like Marie Curie did to the nth degree, I merely came and put a cap over this spot to accumulate the energy of living given off by the sixth side in the form of the worms in the soil, of the roots, of the grasses, of the shrubs and whatever else was growing underneath in this spot of habitation. I never claimed anything on it or that it would do anything, although it's true I would open it every three months or so and measure the energy which was being accumulated underneath. I never claimed it to do anything, but people came there in quite substantial numbers asking could they please sit in the center of it and meditate? Could they lie down on it and spend the night? Preferably could they please sit on a little stump that I had there as a stool? Please could they stay there? And I would always say well, it's up to you. If, when stepping up onto one of the platforms I made twelve inches high, you feel something and if when stepping off you no longer feel it, then whatever you have felt is yours, your interpretation, your feeling. I cannot claim anything. Because Reich had already claimed, or rather people had claimed for him, that the accumulator had done so and so whereupon the Federal Drug Administration came to Farmington and with axes literally chopped up and carried out into the front yard and set on fire the accumulation of boxes that were there. They also burned the books which were written by Reich and which were published in Farmington.

DH: *That's tragic.*

BP: This is a dark history in Maine. And as I travel around the world to different parts and people ask me where I'm from and I say, *Well, I'm from Maine*, most people say, *Where is Maine?* But then I'm impressed by a staggering number who associate Maine not with lobsters or with the rock bottom coast or potatoes or summers or whatever. They say to me, *Look, what did you do to Reich?*

DH: *Well, you've certainly done your piece in righting that wrong.*

BP: In this accusation, they implied that I even to this day, in the great state of Maine, should have known, should have protected him. And anyway he's now gone. For fifty years there are people who have contested this, saying we need his original data. It needs to confirm that he said what he did. And the courts stood by the will and said no. I'm hazy in my mind now as to when the fifty years are up. His argument was that in fifty years people would understand what he did. And it would take fifty years for them to agree. I have physicist friends who have spent considerable time confirming what Reich did and they have asked, *Could we please have his original formulas, his original notes? His sayings as kind of a guideline?* And as I say, I'm a little uncertain as to when the fifty years are up. It's something like thirty-five now. It'll be another fifteen or so before we can have legal access to his original notes.

DH: *That would be in 2004, another fourteen years from now.*

HOLD ON TO YOUR HAT

BP: Yes, something like that. It's shameful actually.

DH: *Oh yeah ... You've told me once that there were certain places in the world that seemed to be particularly strong as natural accumulators.*

BP: Well, I personally claim to have seen orgone energy. I may get it in my backyard, but I remember once while working at the Convair Astronautics (founder of Continental Ballistic Missiles) going weekends down to Tijuana to see the boat-flats. Coming back one day with another physicist, we went through a kind of valley with slopes on the inside. And all the grasses which were growing on either side of the road were aglow with the halo of orgone energy. There's no question about this. It rose above the tops of the growing matter about five or six inches. It was luminous, it was light, it was gray and white. It was a layer of cloud or whatever you want to call it, but Reich called it orgone energy. The energy of living. Furthermore I was in the Arctic Circle between Norway and Russia with Margaret. Again we saw on the slope some vegetation. At the right angle of observation, at the right angle of the light, there it was again. A halo of cloud, a vaporous layer, orgone energy, the energy of living. The human body gives off this energy. Reich's daughter Eva came to Belfast to manifest this effect, namely to run her hands over her father's body, the human body, the nude human body. She followed that spot in the radiation field which was issuing from the point of pain, the point of misadjustment. The doctors in Belfast objected to her presence, claimed she was not a medical doctor [Dick laughing], ran her out of town whereupon she went to Australia to practice her father's ideas. She's now come back to Maine—is in retirement, in obscurity,—still shielding and protecting her father's will, pretty much against her own will, but hopeful like the rest of us that the day will come when we will have the full revelations made by Reich.

DH: *That's an exciting interest too, but it seems so different from your work. As I look at these pieces here I see brilliant colors and a lot of attention to the human form. Does this in any way relate to your Reichian concerns or are they in separate departments?*

Interviews with Bern Porter

BP: No, they're one and the same thing to me. Reich was concerned with living matter, living matter from the trees and from the human body. I was concerned with the light from the human body and its manifestations, its form and space. And all of these manifestations you see here are verbal. If you want another lie, they're really extensions of those original ideas. They're one and the same.

DH: *Because I see all the energy in these pieces and it seems so natural that there should be a continuity with the other concern too. And that's true even with older work of yours, Bern. For example, there was a magazine in the 1940's where you always had a page or a double-page,* Circle *magazine. And you did some very wonderful pages for that. Can you comment on any of this?*

BP: Well, *Circle* magazine at the time followed *View* magazine as being an innovator and a developer and a pioneer. I was associated as assistant editor with a man named George Leite[9]. I left I believe at the end of the eighth issue, and in those eight issues I did covers, I did photo-poems, I did image-grams, I did Mail Art, I did many extensions, many forms, it's true that was for me a rich period. My trouble all these years has been that I have been unable to find a medium to take my sayings.

DH: *Well, you're an intermedia artist like me, Bern! [Dick laughing]*

BP: Yes, yes, and I've always never recovered from when I first found out you would publish me.

DH: *But one of the real beauties of that magazine is that it makes a perfect documentation of what you were doing in the late forties. Because if you look at those pages, some of them feel like they were made much more recently than that.*

BP: Well, yes, that's one of the features that's very satisfying to me, when things that I did forty years ago have as much meaning …

---
9. George Leite (1920-1985): American editor of *Circle* magazine

HOLD ON TO YOUR HAT

DH: *That's fifty years ago! [laughing]*

BP: Have as much meaning today and more meaning than I conceived of. That's very satisfying.

DH: *Yeah. Whatever became of George Leite?*

BP: Well, in a terrible experiment in a chemistry lab he lost his eyesight. And when I knew him he was driving a taxi and his wife was teaching school, and the three of us were putting together *Circle* magazine. I believe and hope he's still living at Walnut Creek. He developed quite an intense friendship with Anaïs Nin while I was developing the same with Henry Miller. And through them we also came in contact with Kenneth Patchen, Kenneth Rexroth, Edward Weston,[10] Ansel Adams, Imogen Cunningham.[11] Those were rich periods, the like of which, I'm sad to say, I do not know are existing anywhere now. I may be wrong. My friend miEKAL [aND] here was showing me magazines today of happenings, events, claims, doings ...

DH: *Oh, there's good people now but that was one of the real nexuses out there. That was in Sausalito, wasn't it?*

BP: Yes, yes.

DH: *Did you have an art gallery at one point?*

BP: Yes, I did, the Contemporary Art Gallery. It was distinguished by being a meeting-place. There were panel discussions, there were exhibits, there were lectures, it was an active center.

DH: *It had a strong concentration on photography too, didn't it?*

---

10. Edward Weston (1886-1958): American photographer and co-founder of Group f/64, the movement's name being a reference to the setting on a camera aperture best-suited for realistic image-quality
11. Imogen Cunningham (1883-1976): American photographer and co-founder of Group f/64

BP: Yes, we touched upon everything. I even had a radio program.

DH: *[laughing] I didn't know that one.*

BP: What was interesting I thought was that the French embassy in San Francisco was besieged by French intellectuals and they in turn wanted the French to be inoculated with American views and would send them over to Sausalito where they would appear in the Gallery. So we had this French connection as it were. [laughing] Quite valuable I'm sure....

DH: *Are there full sets of* Circle *magazine around? Do you have one at Colby?*

BP: A press which is an adjunct of the *New York Times*, the Abbey Press, put it out in book form, in two volumes, and I believe it's still in print.

DH: *Oh good, so those are available now. Well, that's a valuable piece of documentation.*

BP: Yes, yes, very satisfying.

DH: *Of your work which you've done to date, Bern, can you give any sort of estimate of what percent of the publishable pieces have been published?*

BP: Well, just the other day in this transfer of my operations from Maine to California, I counted off thirty manuscripts, thirty projects sitting on the second floor of my home and I wondered what would become of these things. It's true I could send them over to Colby to the Archives. It's interesting that you speak of amiss publications. They took two items of mine, but these two items have sat for something like eleven years at UCLA, just sat there.

HOLD ON TO YOUR HAT

DH: *Well, I know that when Franklin Furnace*[12] *some years ago gave an exhibition of your things, they made contact with the people at UCLA. UCLA didn't really know what they had by you, but they had some amazing things. One of a kind book things which I had seen when I went to the library at UCLA and looked at years ago as Bern suggested. There's wonderful one-of-a-kind books and projects there, all in boxes, very neatly filed away. But the librarians who you were originally in contact with are retired or no longer there. So it's a wonderful collection waiting for someone to do some serious work on. There's a series called the ALO series which refers to the binding cloth that you bound these boxes in, and those are absolutely marvelous.*

BP: About a month ago I wrote to the library at UCLA and said *Look, back in 1942, Lawrence Clark Powell, who was called a librarian, whatever that means, said to me, "We have to feature California people, who do you have?" I said, "Well, I have Henry Miller, I have Janelle Viglini,*[13] *I have myself." He said, "Well, we have to set up you three."* So in 1942 I sent up to Westwood, Henry Miller. And I was told in a letter just a couple months ago that it now represents something like 24,000 labels, of letters and manuscripts, all by and about Henry Miller. I was given the number of catalog boxes and there's something like twenty-four cataloging boxes, and I was told that it's still very active, that people come from all over the world to study from this collection. In fact the third biography of Miller is now being written and will soon be published by a girl named Mary Dearborn.[14] She came to Belfast and spent even more time at the collection at UCLA. I asked them in this letter, *Of the three collections, Miller, Viglini, and Porter, what about mine? Well,* he said, *We're still working on yours. It's not as completely catalogued as we would like. We're in the course of moving but we agreed to complete it at an early date. And as for Viglini, we've never*

---

12. "Franklin Furnace's mission is to present, preserve, interpret, proselytize and advocate on behalf of avant-garde art, especially forms that may be vulnerable due to institutional neglect, their ephemeral nature, or politically unpopular content" [from "Mission Manifesto" at http://www.franklinfurnace.org]
13. Janelle Therese Viglini (b. 1933): San Francisco poet, member of the Institute for Advanced Thinking, Porter's fiancé for a short time, and author of *The Major Filmore Fagan.* See the as-yet unprocessed Janelle Viglini Papers at the Young Research Library, UCLA [http://content.cdlib.org/view?docId=kt5v19p695&doc.view=entire_text]
14. Mary Dearborn. *The Happiest Man Alive: A Biography of Henry Miller.* (Simon & Schuster, 1991)

opened the materials, they're still in the original storage boxes. Of you two, Porter and Viglini, you're more important. We'll finish you and then we'll get to Viglini. But this is quite a problem because Viglini is one of the scholars of the Institute, and I'm her publisher....

DH: *Yeah, well you've certainly made a fantastic statement by putting those things at UCLA. You've got a collection at California, and you've got a somewhat different collection of correspondence with artists and writers in Colby College, and that seems appropriate somehow that you should have a collection on each coast where you've been active. Maybe someday there'll be a collection in the Midwest too, when you've been working with mIEKaL aND and people like that for awhile.*

BP: Well, this archival thing I think is very, very important.

DH: *Yeah, well an archive is a way of stating this is the authentic material. It's not so much that one has to be a beaver for vanished facts or something, but that it's an active working archive for ideas. And that Bern Porter archive has got to be.*

BP: Well, I'm always gratified that people go to these archives at UCLA and Brown and Colby to take notes, to ferret out facts and dates and to testify that there is a contemporary underground of worth.

DH: *Indeed there is. And you just mentioned Brown, do you have work at Brown also?*

BP: Yes, there is. I got my Masters degree in physics at Brown, and in their library there they have maintained not only my original thesis but also quite a substantial number of papers and letters. [James] Schevill, my biographer, has also been adding things to it. All of the notes that he has been using for my biography will go there.

DH: *James Schevill, he used to teach at Brown, a wonderful poet in his own right. He wrote a play in which Bern Porter was a character.*[15]

BP: Yes, it's a fantastic play.

> # Listen to this page.

---

15. James Erwin Schevill (b. 1920): His *Lovecraft's Follies* (1969) was commissioned by Adrian Hall for Trinity Repertory Company in Providence, Rhode Island, 1970. Source: Schevill's *where to go what to do when you are Bern Porter: A Personal Biography* (Tilbury House, 1992)

Photographs by Amy Hufnagel

HOLD ON TO YOUR HAT

Interviews with Bern Porter

HOLD ON TO YOUR HAT

Interviews with Bern Porter

HOLD ON TO YOUR HAT

Interviews with Bern Porter

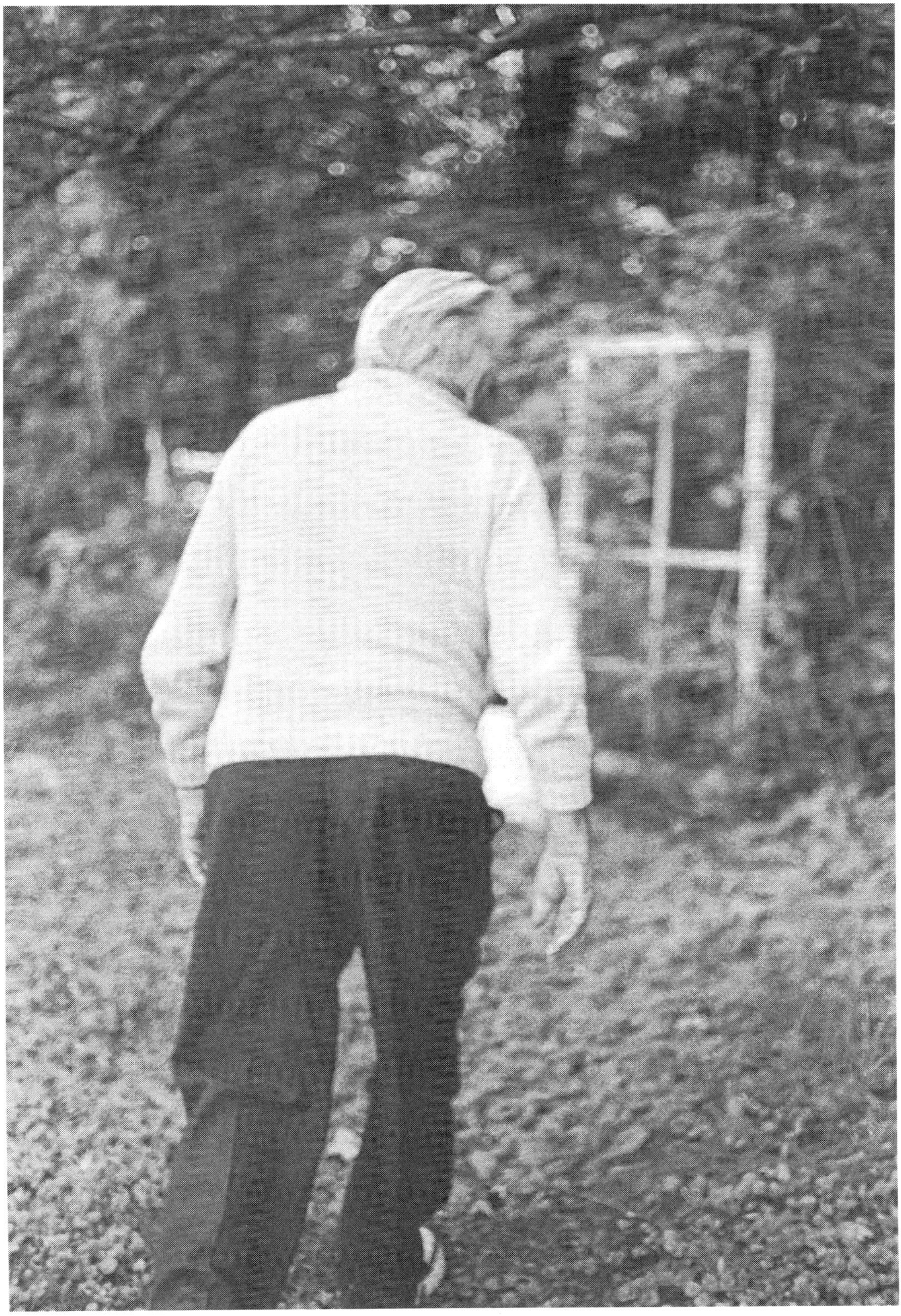

Interview with Bern Porter
by Mark Melnicove
at the Institute for Advanced Thinking, Belfast, Maine, on the occasion of Bern Porter's 80th birthday. February 1991

Bern Porter: Well, downstairs, where you are not supposed to go, there is the radiation.
It's where the nuts and bolts would be
radiated...

Camera man [setting up equipment]: I don't know if I want to risk that.

BP: What do you want? That's
what it is for.

[Camera and sound people fiddle with wires and equipment.]

BP: Well, the day is Friday, we're in Belfast
Maine, the center of international culture.
We are at the Institute . . .

[More fiddling.]

BP: Modern technology is really
something. They rolled out two gadgets in this Gulf
War that just concluded.

They had stuff that had never been tested in the field before.

Some of it was for finding spots below
at night, spots with actual specific spots

HOLD ON TO YOUR HAT

to bomb.
And then the gadgets…

Mark Melnicove: *Can you use them for peaceful purposes?*

BP: We can, but it's set up
so that we can't…

MM: *Could I get one of those gadgets in the car so that when I come to visit you I could pull into the right driveway?*

[Technical difficulty.]

BP: According to me, you and I are supposed
to eat the same food today that the astronauts
did. A pill half-an-inch square for breakfast, and for lunch
another pill, and tonight we'd have
nice paste food.

MM: *And I'm supposed to believe that would keep me healthy and sane?*

BP: Yes, the only hitch is that the astronauts were not exercising, they were sitting in special cultures and special environments, whereas you and I are out and running and …

MM: *Up late …*

BP: Yes, that's true and we're exercising and using
up energy. I've tried to get astronaut food through Brother
Cohen [Maine Senator William Cohen] and he's sent me
some substitutes, but he said you can upset the economy

Interviews with Bern Porter

with real astronaut food, as we can't upset the spoon bouillon
people, and those who make napkins, chairs, candles, and tablecloths.
You can't do this.

Whereas
At night I want a choice of paste,
I want either strawberry or pineapple or lemon.

MM: *And you would like to take this through your mouth rather than with a needle?*

BP: I'd prefer…

[technical difficulty and soundchecks follow]

BP: Belfast Maine, culture capital of the world.
Here we are.
Hello, hello, hello.
So here we are, in Belfast, Maine.

Am I turned on or what?

Now can you hear me there?
Hello. Hello. Hello.

OK, OK. Testing Belfast, Maine.

[static]

Sound Woman: Bern, you're static-y, but you're fine.

BP: OK, OK. Now you are getting me?
You want me to speak up loudly?

HOLD ON TO YOUR HAT

O loud enough?
Alright!
This is the world headquarters of the oldest and largest institute in the world. It started off with Caesar when he invaded Gaul; his ideas were mixed in with Gaul; then England, the druids.
The ideas came to shores of Maine around
the year 1000.

MM: *Were you here then?*

BP: In the spirit and energy I was here
then, yes.
I want to make it clear that I have always
been here, and I will never
leave here, the energy
in me.

Which I will explain later from the point of view of physics.

Around the year 1000 these ideas were
Brought, by what turned out to be
fishermen, who in the course of fishing along
this coast, stored salt in Belfast,
Maine.
They needed the salt for the storage
of the fish taken to
Europe.
My grandparents came from the borderline
between Scotland and Ireland.
They came up the St. John River to the point
in New Brunswick called Woodstock.
They crossed over to a branch of the Kenduskeag

Interviews with Bern Porter

River, built a cabin ...

MM: *What year are we talking about?*

BP: We're talking about
the year 1830, when the Institute was
reorganized by my grandparents who
established Porter Settlement and who started
off in a log cabin.

When I was born in Porter Settlement there were 23
Porters.
With uncles, cousins, grandmothers, the entire
relative field.
When I was there last there were only
two Porters by marriage,
two by blood.

Here I am faced with traumatic
situation of the Porter
Empire collapsing with someone coming up
from Newton, Mass. and the Porter line
exhausted, run out

and someone by the name of Smith taking over the Porter Empire.

To come back to the Institute: it was reorganized in
1830, it has been in Maine almost
a thousand years, the ideas and the principles
which we use ...

HOLD ON TO YOUR HAT

MM: *What are some of the ideas and principles?*

BP: The principles that we use? Someone
has asked me what is advanced thinking
and the correct answer is whatever is
obvious, natural, but mostly
whatever is obvious.

And our culture is setup to overlook what is
obvious and what is
reasonable and sustainable and what is
true and right.
We are fed on hypocrisy and greed and money so the whole
thing is distorted, the things we concern
ourselves with are the natural
state as it comes to us from
nature itself.
Physics is the study of nature, so
we study nature.

There are people who like to make
a joke of this. They say
what do you guys think about?
So to make it easy we say
on Monday, Wednesday, and Friday we think
about sex.
And on Tuesday, Thursday, and Saturday we think
about money.
And on Sunday we go
to church.

And everyone thinks this is sufficient.

Interviews with Bern Porter

When the true answer is
we are concerned with the natural
forces around us and what
is obvious.

[Fuzzy sound, stop-gap measures to fix.]

BP: From Einstein, we have the idea that
as we approach a membrane and we push it
at a certain point, another point comes out and hits
us in the face.

And in this membrane are swinging windows through which
we may or may not be
able to pass.
There are gurus in certain cultures who sit
under trees for 40 years with the hope that in
the swinging of the windows they can get
through. And if they get through they become
God, which

is impossible.

But we here at the Institute are aware
that the membrane is there.
We cannot go over the top of it
or under it.
We cannot go to left
or right.
We only press it forward. We are aware
of the swinging windows and we endeavor to interpret
the energies that come through

HOLD ON TO YOUR HAT

to us. These energies are of complete
complexity.
One was called by Wilhelm Reich orgone
energy. There are
others: laser energies, radiation energies, radiation
from electrical, there are many.

MM: *Can you show us one of these membranes? Is there a membrane in the room with us?*

BP: Yes, there is
a membrane and the swinging
windows are here with
us.

Could I
point out that at birth we
are endowed with five
senses.
Naturally.
Could I
point out that it's possible
to proceed from the 5th
sense to the 6th, 7th, and 8th?
There are gurus
in religions and cultures who aspire
and claim to have done
this. The horrible fact
is that our hearing is
gone; we have hearing aids
to help us; our eyes are gone,

Interviews with Bern Porter

in 200 years biologically speaking we will be born without the capacity to taste or smell.

So we have mechanical
gadgets propping up these five
senses, so the individual who can
function 3 or 4 seconds out of 24
hours using all five
senses is impossible.

We are so bruised
with the food we eat and the atmosphere
around us and with our habits
and whatever,
even with our clothes.

We have ruined the five senses with which we have been endowed.

MM: *And when did you first become aware of this?*

BP: Well,
all of this I'm speaking
of, I think
is obvious.

I'm saying that the person who uses all five senses is rare indeed.

And we aspire
here at the Institute to do
that.

HOLD ON TO YOUR HAT

MM: *And what are some of the ways you aspire to do that?*

BP: With periods of rest.
We object to the vitamins
which are sold and special
diets that are compounded.

We are aware of the horrible
fact of food first being cooked, then
frozen, then unfrozen, and then partially
heated, and then
consumed.
As physicists

we see the molecular
structure and value of the food being
completely destroyed. And yet
this is a common
way of diet for
American people.

MM: *But just a while back you were lamenting the fact that we didn't eat like the astronauts by consuming their pills and paste.*

BP: I am very sad
about this, particularly
that pill, only ½-inch
square; you just nibble
on the corners of it, it
gives you all the energy

Interviews with Bern Porter

your body provides
assuming you're going to be sitting in a special chair and not exercising.

MM: *What kind of food did you eat when you were growing up in Houlton, Maine, in Aroostook County?*

BP: I was brought up in the Potato
Country, where we had fried
potatoes,
potato cookies,
potato pancakes,
potato cakes,
potatoes,
potatoes,
potatoes,

and on Sundays, a special day, we had a baked potato.

And the result is 40
years later my brother presents
me with a lobster and by merely touching
this my face turns green and
yellow and I
collapse and I've
been poisoned.

It's not possible for me to have anything out of the sea

because I'm from the interior, I'm from
the land of potatoes, and my
body is a machine, it's been
adjusted to a certain

HOLD ON TO YOUR HAT

way of intake, and a certain
way of life.

[technical break]

BP: I claim to be
an Abnaki because
the Abnakis were here ten thousand
years before the Europeans
arrived in the year 1000.

It's important for you people to know
that this spot where we now are is
710 feet above the low
mean tide.
This spot,

Belfast Heights, is highly
charged with the energies of these
people, with what's called
the energy of habitation.

It's unfortunate that there are telephone
lines and electrical
lines on the street that interfere
with this, but
this spot
and the basement
and the accumulator
in the back are highly

enriched by the energy of habitation of these people.

Interviews with Bern Porter

People give off
energy, people have
plasmas and they give
off energy, and this
energy inhabits the soil, trees,
shrubbery, and it is
of a very high
potential
value in this particular spot and in this house.

And that's the reason the Institute is here.

Photographs by Read Brugger

Interviews with Bern Porter

HOLD ON TO YOUR HAT

Interviews with Bern Porter

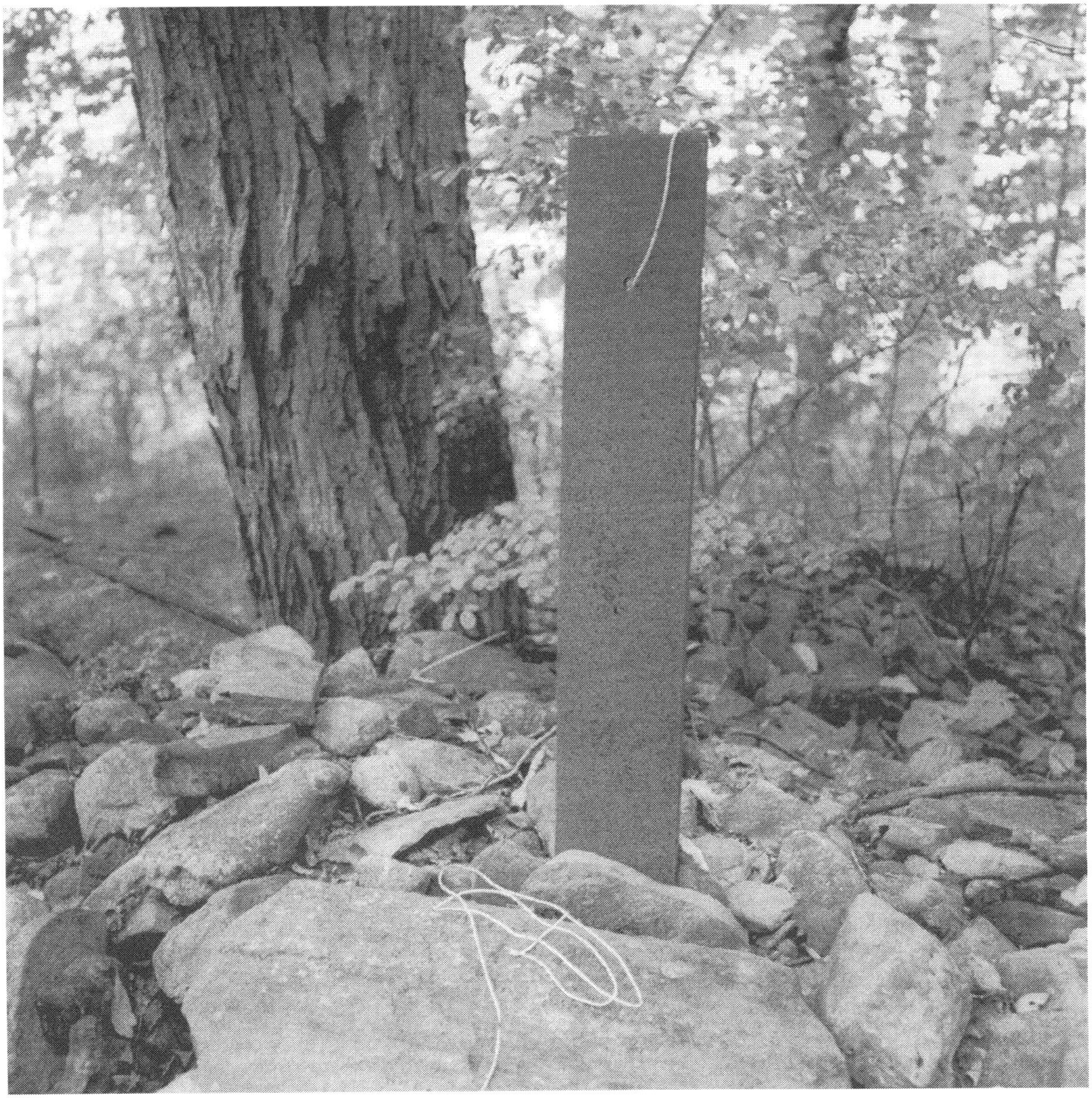

Interviews with Bern Porter

HOLD ON TO YOUR HAT

Bern Porter interviewed by Phobrek Hei and Sasha K.
at his home in Belfast, Maine Saturday, August 11th, 1990

Sasha K: *Did you become interested in literature at Colby College?*

Bern Porter: People are victims of circumstance. In my case, my parents didn't have enough money for a place like Colby, so I spent my first year at a junior college. However, I then received a scholarship and was able to attend Colby as a sophomore. During that year I had a job tending the furnace at the men's gym for meals at the women's dorm. But by the next year, I had become a lab instructor at Colby for fifteen dollars a week because I happened to be good at physics. By inclination I might have studied English, philosophy, not physics. But I was good at it, so I even ended up going to Brown for graduate studies. Now, while I was in physics, I had a hobby—writing poetry, literature—with the result of my being the author of eighty-four books and the publisher of over four hundred. I started in 1929 and have been going for sixty-one years. I've managed to manipulate my circumstances through my hobby—writing and publishing.

Phobrek Hai: *You were the first to publish Henry Miller in the United States, weren't you?*

BP: Yes, yes. In the course of traveling to Oak Ridge while under surveillance, observation, etc., by the government, I made my way to L.A.—to Henry's place at Westwood. And in effect I said to him, *Look Henry, I have some money as a physicist, and I'm interested in your writing. Do you have any manuscripts?* And Henry said, *More than I know what to do with.* So I ended up publishing seventeen of his, the first being *What Are You Going to Do About Alf.* I guess I sold it for about $1.50. But after I'd published it, someone called my attention to the four-letter words and such, and said, *You're a fine Colby graduate, how could you do this?* So I took a marker and crossed out all the four-letter words in eleven copies of the book. I think now copies with those marks in them are being sold at eight to twelve hundred dollars.

HOLD ON TO YOUR HAT

SK: *You were also an acquaintance of Gertrude Stein?*

BP: Yes. Gertrude was a San Francisco girl, money in the family, and she wrote in her style—repetitious. People called her work meaningless, called her a crackpot, which was not true at all, because by expressing herself through repetition, image after image, she could build a structure.

In 1937, I went over to Paris on the Normandy (this was near the World Exposition in 1937), and a lot of Americans were coming by to see her at her home there saying, *Gertrude, you're far out.* Back then, her house was like a bus station. She was one of three there who created their own whole philosophy—Stein, James Joyce, and Abraham Lincoln Gillespie, who went on from where the other two left off. And now they're gone. Forgotten. This is one of the great sorrows of our culture. People like these die, are forgotten, and then are resurrected twenty years later. Henry Miller, Kenneth Patchen ... twenty years later the critics finally begin to realize their meaning and they say, *Well, you're dead now, but sorry about kicking you around when you were alive.*

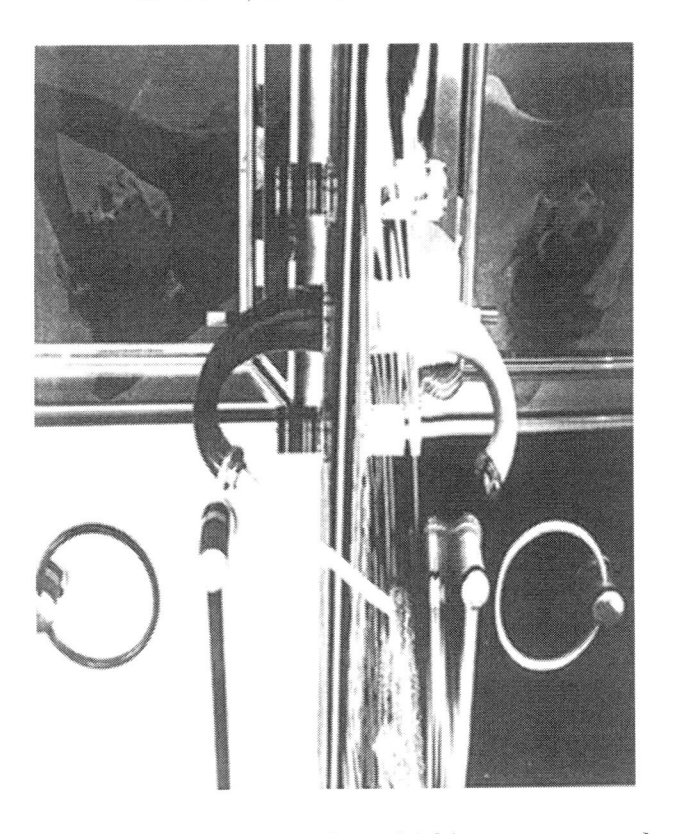

Forty years ago, Anaïs Nin developed what is now the feminist movement. In effect, she was saying to women, you were born to create and not to clean. In the eighth grade, this woman walked out of school for the last time. But she continued to get up every morning and go to the library instead. She proceeded to read every book, from the top to bottom of each shelf. Every book in sight she read. Meanwhile, she was writing her diary, which I read when I was wondering, *Who can help me?*

Interviews with Bern Porter

SK: *And after coming back from Paris you were drafted into the Manhattan Project?*

BP: One day I was called before the draft board. Luckily, one drafter knew what physics was. So instead of running around in trenches with a gun, I was told to report to Princeton. I was interviewed and accepted. I was shipped to the Princeton Physics Department, where I worked with some grad students under me, testing and separating isotopes. In those days, we didn't know what uranium was, so the janitor would just come sweep the stuff up. Nowadays, they have special suits to wear, mechanical hands, and a wall separating you from the uranium—only then you can touch it. Back then we were breathing it. I have been irradiated. Every four years I go to MIT, where I'm examined by other physicists tracking the elements of radioactive material that went into my body. I cannot contaminate anyone else. They send me vials before I come—one I breathe ten breaths into, the other I am to inhale ten from. And this is all measured to see how much radioactivity I am breathing in and out. I am, in effect, a living specimen of one who has been irradiated. My master's at Brown was on uranium, and now I've got it in me. When and where and for how long I was exposed is unknown. The result being—I've given my body to physics. When I die, all of my organs will be used for experiments, and my skeleton will be crushed into powder. On examination of this powder, I will end up as a point on a graph, with points of other persons. The point defining me is a result of not drinking, not smoking (clean living, whatever), heredity, and personal characteristics—all ends up on this one point. The sad thing to me is that what they know of me from all this will not help you at

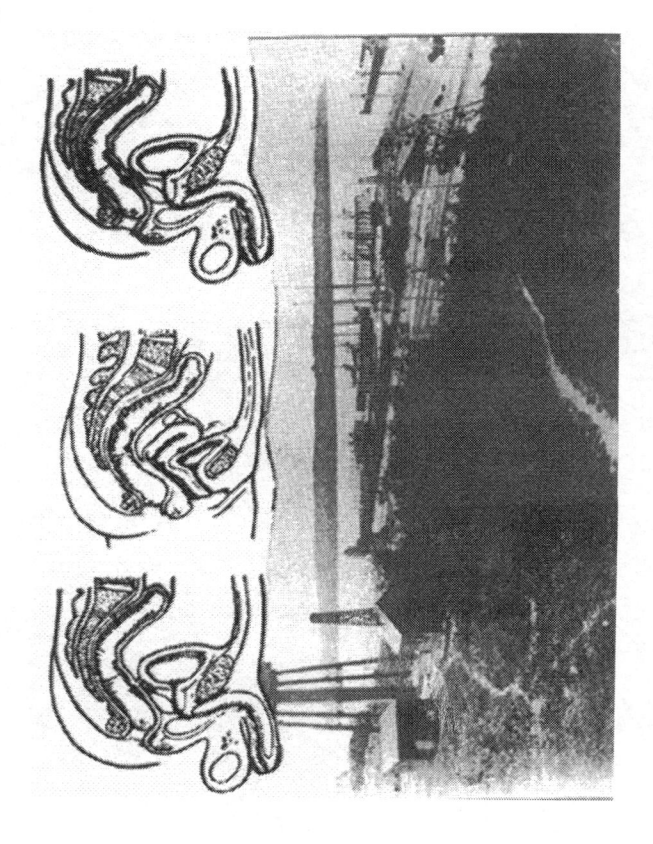

HOLD ON TO YOUR HAT

all. You are sitting here and being irradiated by the atmosphere.

Three and a half years after the explosion, I had the experience of walking through the streets of Hiroshima, and a little kid came up to me with something that looked like a golf ball, and he wanted to sell it to me for fifty cents. What was it? What did it used to be? The boy told me—it turned out that this had been a milk bottle. The radiation and heat had reduced this bottle to its simplest form, a sphere. A sphere made of glass. Having figured this out, I thanked him.

A man came up to me. He was offering to show me his back for fifty cents. Well, I told him that I'd like to see his back, but that I wouldn't pay to see it, not with him capitalizing on this. So right on the street, he lifted his shirt and showed me. It was blistered, pimply, blown up, red, white, yellow, incredible. Doesn't this pain? I asked him, and he said, *Yes, it does, particularly when I move.* I asked him how he slept, and he said that he had to sleep on his face all the time.

So we had these two people on the street capitalizing on their misfortunes. And then a man came up to me, and he said *Would you like to see my sister for fifty cents?* He led me down an alley to a typical Japanese house and there in the middle of the floor was this man's sister. There is no way to describe ... her eyebrows were gone ... her jaw was ... the lower part was gone. She was horribly blistered, like the man on the street. I said to her, *Where were you when the explosion took place? I mean, under a roof? Near a wall?* Was she dressed, not dressed? She was taken to America, as she had agreed to come and be their guinea pig for her physical restoration. Obviously her eyebrows couldn't grow back and neither could her hair, but she could wear a wig, etc.

Sometimes, in Hiroshima, I would go into a bar and I'd be asked, *Why did you hit us? We didn't do anything. We're a peaceful city. We didn't make the bombs. We mind our own business. Why did you hit us?* And when I replied with, *What about your attack on Pearl Harbor?* They said that the Emperor told them to fight, so they had to. If the Emperor says go, you go.

Then they said to me, *Why didn't Einstein come over and meet with our Emperor—he's a biologist of international acclaim—and sit down and explain to him what it was, what it could do?* Well, we didn't do that. *And if they couldn't get over here, why didn't they give a demonstration? Like show the island five miles off being hit?* Which we could have done, incidentally. But war is hell. So, here I am, a frustrated person, in a sense.

In the Institute of Advanced Thinking I have developed the fusion of Physics and Humanities and have been exploring that fusion. You can combine various fields for certain effects: say, Physics and Architecture; Science and Sculpture; Science and Literature; Science and Art (which is the most powerful one). These are developments which I created ... my contribution to culture.

So the Institute in Belfast has thirty-four scholars around the world. Members cannot be employed by a university, cannot be teachers, cannot work in a library or some sort of institution. A member must be a freelance physicist. Ideas are the work of a movement. The results are not copyrighted or patented, they're all free. I visit these people four months out of the year. And occasionally scholars from Germany and Japan come here. This is the headquarters for the Institute for Advanced Thinking. It is based solely on the fact that I had taken part in the atom bomb. I want to show that physics can do other things than blow up the world. First as a physicist, and second as a publisher, I combine the two—fusing physics and humanities.

SK: *In the direction you see society going, do you think that this fusion can happen?*

BP: It's a struggle. The press called us a think tank and in the course of our thinking, ahead of our own time. While the government is subsidizing tobacco crops, chemical warfare, etc. It's not easy to go to people and tell them that physics can save the world. Fighting with physics and fighting with literature ... Some people say, *Why bother?* Well, I can't do anything else.

SK: *Similar to Wilhelm Reich?*

BP: I have the only horizontal orgone energy accumulator in the world in my backyard. However, it depends on a nearby waterflow, which was recently destroyed by some construction going on in the neighborhood.

Yes, I knew Reich and admired his work. It was very sad that he was put in prison and left to die there. Looking at how such people are treated, it makes one more pessimistic. In the street, people point to me and say, *Hey, there's that crackpot*. There's nothing I can do to prove my sanity. Just thinking about it makes me sad. I'm seeing things gloomily, but here and there are little holes—like the Institute.

Reich came up here from New York, to Maine. He came because of the pure air and water here (or, it was pure in those days). He came up to Rangeley—a perfect place which you should go to on some weekend. A perfect site, mountains and water and such, where he was able to do his experiments on orgone energy.

It was very sad. In his will he said his notes couldn't be opened for fifty years, and there's still about twenty more to go. Meanwhile, there are people who want to get his notes and equations, to see what he said. But in his will he said that it would take fifty years for people to be in a position to understand him. Even when people like me are ready, needing his original works. Desperate.

I took measurements for Reich for eleven years—the accumulations—though I've lost it all. The neighbors next door cut off the flow of water coming to my property, making it back up and dry out. All my accumulations, my measurements, depended on the flow of water. It broke down. Eleven years of work gone.

SK: *The government is trying to shut these people down—scientists like Reich, writers like Miller or William Burroughs. Why?*

BP: Society doesn't want original ideas or a rocking of the boat, an upsetting of the waters. Miller was a dirty old man, they say. Take *Murder the Murderers*, which I also published, which is what the title says—kill the killers. They killed in Panama, now in Kuwait. We specialize in wars. It was a nice war. Well, someone has to make the guns, planes, boats. It's good for employment.

So it's not easy. When you criticize the government, which is supposed to build roads, schools, and hospitals, but instead becomes a killer, you're accused of being a crackpot. *Why don't you be normal, stop fooling around?*

SK: *Do you sense any sort of conspiracy within the setup of society?*

BP: The whole system is set up to make money at any cost. Destruction, herbicide, pesticide, Star Wars—all based on making money. There is no particular concern for the consumer. I see a pattern, but not a conspiracy. Unintentionally contrived. They think they're helping make a better place, a place where we're not indoctrinated to use gadgetry and electronics to get along. Not conspiracy. But the way it's set up, which we accept without thinking, is yet harmful and disastrous. But what can I do? Reich, Stein, Miller, we're all seen as crackpots.

PH: *And yet, you do continue to make the effort to express your views. Could you give us some background on your art?*

BP: I am the inventor of Found Art. This was a tremendous development for me. I was born without anything and therefore had no money for an artist's tools such as easels, paints, brushes, etc. So, at the age of four, I invented Found Art. I would go to the Post Office wastebasket and go through it, pulling out things that in some way would interest me. I'd find printed words, word gems, and combinations of words that were poems in themselves. With a found poem, I would go through the wastebasket, take scissors and glue and put it together. Cut and reassemble it. I am the inventor of the whole philosophy of Found Art.

A problem with being seen as a crackpot is that I'm so old that I can't get a publisher. It takes a very unusual one to want to print my work because they call it avant-garde, unusual, experimental. *You can't do that—why bother?* Because I can't do anything else. Found Art will never be recognized as a true art form in the United States. Not in my lifetime. And why not? Because we are crackpots. Half-crazy. Avant-garde. Branded. I've been branded for years.

SK: *Can you talk about* The Last Acts of Saint Fuck You?

BP: There is a figure on his deathbed. A crackpot, an avant-garde. And in this book he is detailing what is to be done to avenge his mistreatment from society through his last twenty-six acts, which are written alphabetically. When someone says, fuck you, this is what in effect they are saying.

It's wide open. Just keep going and do what you must do.

CROP

Interview by Mail with Bern Porter
Seth Tisue and Brad Russell
(1991/1992)

Seth Tisue / Brad Russell: *What is the connection between Bern Porter the physicist and philosopher, and Bern Porter the found poet? (How are your other concerns reflected in your poetry?)*

Bern Porter: Creativity of a concentrated high order is involved in physics, philosophy, poetry and in common to all three activities. My life has been deeply involved in these areas and continues so daily today (at age eighty-one).

ST/BR: *What is your attitude towards the "found" materials in your books? Horror–bemusement–detachment–wonder? Is your work a condemnation or a celebration? How do you want your books to affect people?*

BP: My most devoted fans range from eight to twelve years of age. Laughter and amazement engulf them because they are not yet contaminated by religion, government, education, culture, parents.

Anyone older than twelve does not readily respond to my thesis of showing what I have found, not what I am looking for.

The pages are unnumbered, there is no plot, no description, no thread of events, no dialogue, no characters.

The volumes are best opened at random for a two to four page visual response, then put aside. A year could elapse to embrace everything.

The total intent and impact is to amuse and induce thinking.

ST/BR: *You have spoken of "escaping" from contemporary culture, but isn't your found poetry an immersion in it?*

BP: The found and prepared found materials are cut directly from all forms of contemporary culture available in vast quantities in uncopyrighted printed matter.

HOLD ON TO YOUR HAT

The elements presented are then attacked, mocked, scoffed at, diminished, slandered, libeled.

I personally live all of this out daily in my life finding myth, illusion, falsity on all sides forcing me into an escape pattern.

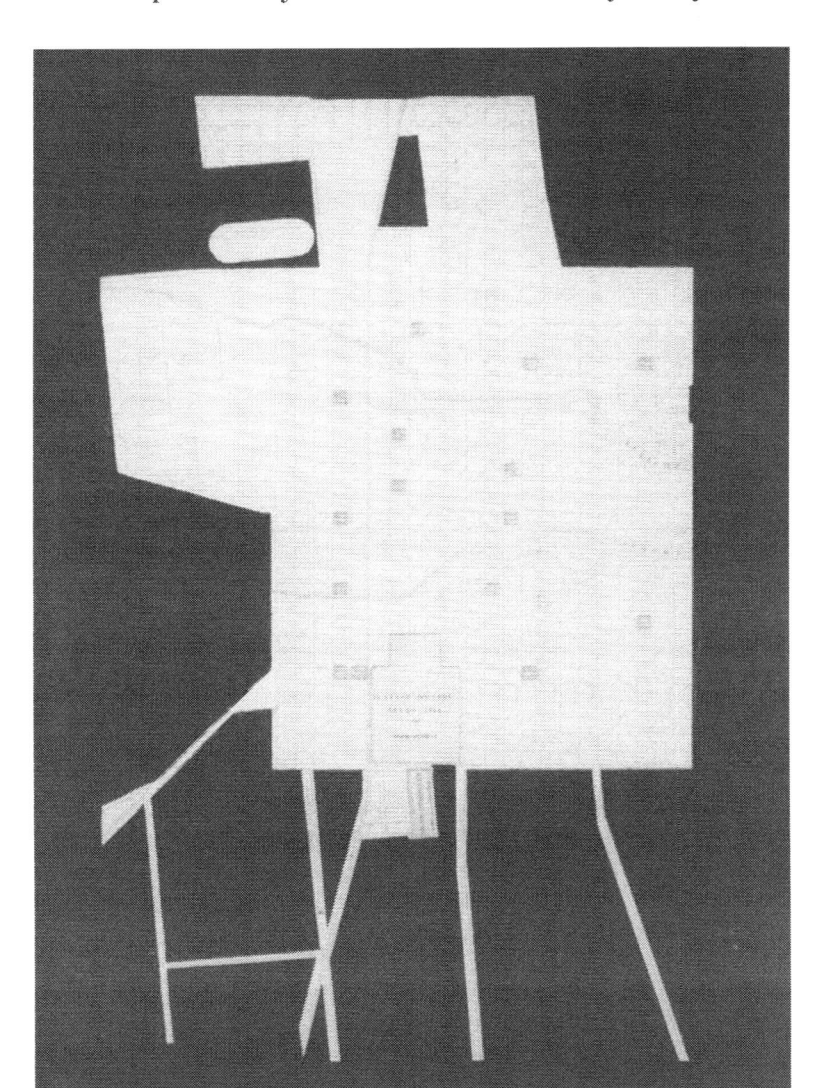

ST/BR: *Many of your recent "founds" are graphic rather than textual—glossy magazine pictures rather than the checklists and catalog pages of earlier work. Why?*

BP: It is interesting that few people actually read anymore but depend solely on visual matter to convey meaning. Publishers and editors and designers sense this trend and supply now more photographic imagery than words. Thus my early sources have been drastically removed.

ST/BR: *Describe the Institute of Advanced Thinking. What kind of research does the Institute conduct? Does it accept students? How many, and on what criteria?*

BP: The Institute is involved twenty-four hours a day, seven days a week, in the fusion of physics and the humanities.

Interviews with Bern Porter

Thus the union of physics and art becomes SCIART, physics and architecture SCIARCH, and so on through seven forms.

It operates worldwide through a network of scholars concerned with all forms of energy, matter and anti-matter.

Students are invited to describe their interest on a single sheet of paper. If physics can be or is involved in any fashion they are invited to come to a five acre site with appropriate living and working spaces in Belfast.

ST/BR: *What are the rest of the "seven forms?"*

BP: The full roster of forms resulting from the fusion of physics with the humanities (first called Physics for Tomorrow) is:

    SCIART, the union of physics and art
    SCIARCH, the union of physics and architecture
    SCIED, the union of physics and education
    SCIMUS, the union of physics and music
    SCISCU, the union of physics and sculpture
    SCICOM, the union of physics and communication
    SCIENG, the union of physics and engineering

People often ask *Why SCI when physics, the base of it all, could be reduced to PHY?* The answer is that, aside from phonetics and appearance, there are many forms of science and all of them involve physics.

It is interesting that translators into other languages have accepted them as is, being unable to reduce them to foreign language forms. Not being copyrighted or patented, the words are now in common use worldwide.

Of the seven forms SCIART and SCIED have substantially come into their own and wide acceptance.

SCIART will be fully explained and featured in the 256-page (fifty illustrations) personal biography of me by James Schevill, to be released in both hard cover and paperback by Tilbury Gardiner this August (1992).

The Community College of Maine announced its formation as a functioning public service on January 15, 1992 with a catalogue covering sixty-four interactive television sites in Maine and providing educational (SCIED) courses transmitted for students to see and hear instructors on TV monitors. The Faculty will teach from a common campus classroom with one electronic camera (SCIED) focusing on the instructor and another focusing on charts, diagrams and other visual materials. A talk-back system permits students to interact with the instructor by cordless telephone.

After forty years of development an Institute vision has been truly realized.

ST/BR: *What are your achievements as a physicist? What are your achievements as a writer and artist?*

BP: *Who's Who* (in the West, in the East, in Space) would cite my accomplishments. *Contemporary Authors* would reference my written work.

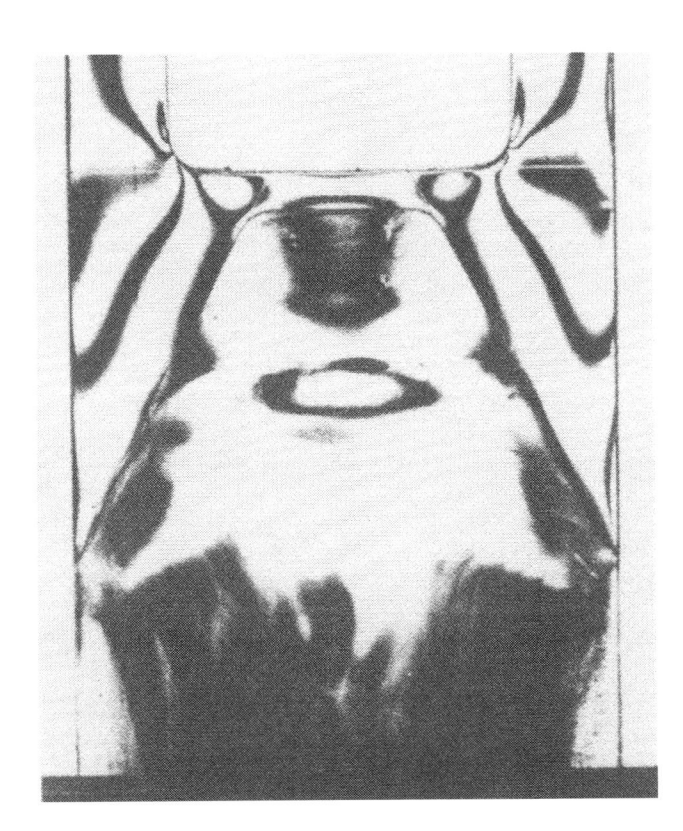

ST/BR: *Can order be created, or only discovered?*

BP: Order can neither be created nor discovered. It can only be examined with the advance understanding that its ultimate truth cannot be known or fully revealed.

ST/BR: *What is the relationship between the intellectual and the physical? Culture and nature? Are these useful categories?*

Interviews with Bern Porter

BP: Our culture has set up categories for cataloguing, indexing, describing, referencing. These are crude, misleading and confusing. As such they are responsible for many ills including the impossibility of total communication.

ST/BR: *Should one withdraw from society? Attack it? Work within it? Ignore it?*

BP: One can only with forced difficulty adjust to it, sadly with mainly unsatisfactory results. At best one can only bungle through.

ST/BR: *Explain your theory of "plasma."*

BP: A minister tells you you have a soul.
A chemist tells you you are ninety-three percent
      water and seven percent minerals.
A research physicist, I tell you you are
a plasma of energy.
This energy was/is ever present from and
      before time began.
It sends, it receives.
It expands, it contracts.
It lives, it dies.
It vibrates, it remains passive.
It cannot bleed, be bruised,
be cut, be dissolved.
Humans are unable to conceive of
it, multiple simultaneous
actions.
It is non-solid, non-gaseous, non-
liquid, invisible.
Being ever present it was crystallized in an instant when
your father/mother conceived.
Nine months later it was given your name.

HOLD ON TO YOUR HAT

When you die, your body—its container—falls
away as it goes on four-dimensional
through time, time through it in a straight
line of an arc through space.
It never dies.
It returns its radiation to our memory.

ST/BR: *You have said that one of the sorrows about Reich is that it'll take another fifty years or so to confirm his work. What do you feel has been confirmed and what remains to be confirmed?*

BP: So far we have only touched his ideas in a partial attempt to believe, to confirm, to duplicate, to extend his ideas. Much now depends on our having his hand-written equations (removed at his request from his jail cell and now sealed by his will for fifty years).

   It is hoped we can then duplicate his total findings.

ST/BR: *What has been your personal experience with orgone energy?*

BP: For twenty-five to forty years I have approached it under varying circumstances.

   With physicist Ralph Markson I saw it in a valley on a trip home to San Diego from Tijuana while working on intercontinental missiles for Convair Astronautics. With my wife Margaret on a journey above the Arctic Circle, near the border of Sweden and Russia, I saw it again.

   After seventeen years of experimental use and measurement here in Belfast, I was obliged to dismantle the only known horizontal orgone accumulator in the world. May 16, 1991.

   Its water-field had both been misdirected and evaporated, its structural components decayed.

   As such, the physics department of the Institute of Advanced Thinking had been inundated with great permanent losses. The world's oldest and largest horizontal orgone accumulator and seventeen years of its recorded measurements, as I said, have long gone. More fatal are the years that concerned workers throughout the world have been waiting for the public release of the much needed original formulae Reich passed out of his prison cell via scribbled notes on scrap paper and a trusted female friend. Attempts by his daughter Eve to have the courts break his fifty year-old will requirement, perhaps set by him as fifty years for the rest of the world to catch up with him, have failed. Many experiments have stalled awaiting these final notes.

    Meanwhile, blankets made to his specifications of alternate layers of fiber and metal can be obtained on the open market with the maker/sellers making no specific claims of health benefits resulting or not resulting.

ST/BR: *Even some supporters of Wilhelm Reich think he went crazy late in life. Did you know him during this period? Was he insane, or did his thoughts just leave everyone else behind?*

BP: He was merely fifty to a hundred years ahead of everyone. Not insane, his ideas, truths, experiments will be verified in full.

ST/BR: *What do you think of other individuals that, like Reich, are usually branded as "pseudoscientists," such as Fort, Velikovsky, von Daniken, etc?*

BP: Such words make an easy way out when one does not believe.
    Better to be cautious, patient, wait and see.
    In 1992 many classical concepts are under evaluation.

ST/ BR: *As a physicist you participated in the development both of television and the atomic bomb. What sort of future did you see for these inventions at the time? How do you feel now about how they were and are being used?*

BP: The current spectacle of the
prevailing scene is beyond
description.

Television,
the boob tube of the
masses, consumes hours
with drivel, mindless
and a loss.
The bomb has
developed into an
uncontrolled monster
called atomic energy
which takes much
of my time decrying
dump sites for the
waste falsely labeled
"low-level."

Man with his
gadgets have
befouled the
rest while I
sincerely intended
the exact opposite and
daily work to that goal.

ST/BR: *Describe, in as much detail as you're willing to give, a few interesting past or present research projects of the Institute?*

BP: After our great satisfaction arising from the use of SCIED here in Maine, we are striving daily to advance SCIMUS and SCIARCH to a comparable position.
    From May to November of every year the Institute features outdoor sculpture shows and indoor gallery art exhibits to which visitors are invited to bring tents, sleeping bags and swim suits. Some stay for weeks and generate new approaches to old ideas.

Interviews with Bern Porter

ST/BR: *What have you discovered on your travels? What can primitive cultures teach us? Us them?*

BP: No one can call himself/herself educated until he/she has traveled extensively world-wide.

Now eighty, I started when I was nineteen, visiting cultures 2,000 to 50,000 years old in order to prepare myself and my obligations as a physicist.

In this sixty-year operation I learned that the 350 year-old U.S.A. is not only backward and undeveloped but has a long way to go.

About all we have to offer is a Constitution, a Bill of Rights, apple pie á la mode, a flush toilet and some medicine.

The cultures I saw maybe could use some of our medicines with the understanding that most of theirs from nature are probably better. Some may never get a Constitution or Bill of Rights. All have gotten along without apple pie á la mode. Some without flush toilets. Thus Americans by comparison do not have much yet to offer.

The American procedure for years has been to send missionaries to the outside (or third world, as it is called). All religions have been doing this, followed by businessmen.

The process should be reversed at once with people of the world coming here to teach us how to live, care for the elderly, educate the young, live with nature, be humans.

We have nothing to teach them; they have much to show, to teach us.

As I said, we may have some medicines.

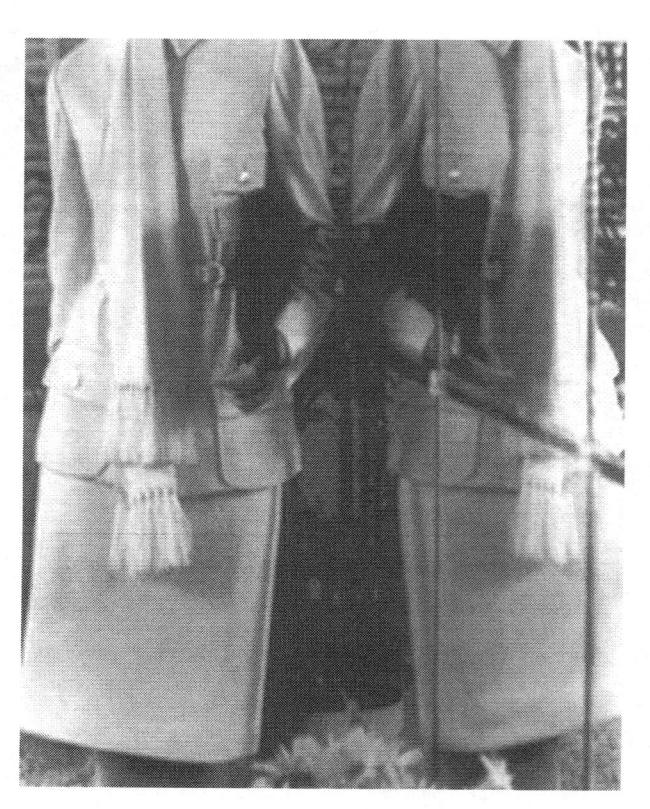

HOLD ON TO YOUR HAT

ST/BR: *What signs of hope do you see for the world? Are the last 5,000 years of human history salvageable, or should we just start over?*

BP: For the last 5,000 years humans have botched clumsily, awkwardly along hatching up technologies which by 1992 have more than fouled the human nest worldwide to a point of unliveability in many areas. More, humans enjoy the mess they have created beyond the point of doing anything about it, that is if they could.

The plasma of humans is such that most will bungle through, but sadly botanical specimens, vegetation, birds of the air, fish of the sea, insects and small animals will not.

The oil fires in Kuwait are far from aiding the process.

Bungling mankind will bungle through.

ST/BR: *Are fundamentally new approaches to the problem of food, clothing, shelter, etc. possible? Do you know of any?*

BP: Considering the piles and piles of money spent on the research, development, manufacture, testing, and use of cubes and pastes for the astronauts, every man, woman, and child in the world should have them free and at profitless cost from this date forever onward.

The days of knives, forks, spoons, napkins, cups, saucers, plates, tables, chairs, and can openers are over.

Clothing is to be one piece, unisexual, disposable, without zippers, buttons, or style changes every six months.

The clothing becomes the shelter.

ST/BR: *Future plans?*

BP: The Institute of Advanced Thinking, now spanning almost a thousand years in Maine and incorporating fifty-thousand years of human creation, will occupy me until, on passing, what human parts of use will be given away and my skeleton ground to dust for the study of its radioactive content by inquiring physicists concerned about the future of humanity.

The thirty-four scholars world-wide of the Institute will carry on.

Five publishers now turning out my four books a year will prevail until then when rare books dealers will make them available for a hundred dollars and up per each.

The eighty to ninety books of my authorship spread throughout the English-speaking world are my children. May they multiply and enrich.

Recently through a children's search service I have adopted an eight year-old child named Jeyalakshmi. She lives in Lakshminarayanapuram, India with her mother Mariammal and her father Kali.

She is eight years old, has black eyes, black hair and a brownish complexion.

Her native tongue is Tamil. She is in the fourth grade and with time will carry on the work that has been established here. And this in spite of and because I have been without legal help representing myself from a sitting position in the local soup kitchen at both District and Superior courts in twenty law suits since 1987 to recover my life's savings from a local con artist and entrepreneur.

The five publishers–principally Tilbury of Gardner, Maine–will continue to represent me along with Cong of Verona, Italy and Volatile of Cincinnati, Ohio.

Bern Porter interviewed by Judith Hoffberg

Editor's Note: This is a reprint of an interview with the late Bern Porter, printed in *Umbrella* 1:5 (September 1980). During a brief visit to Los Angeles, Bern spent an evening in Venice during which time Judith Hoffberg, the editor of *Umbrella*, interviewed him on January 22, 1980. Poet, physicist (he helped develop the cathode ray tube, pioneering television as well as the atomic bomb) and innovative bookmaker, Bern Porter lived in Belfast, Maine when he wasn't going around the world on cruises. He died in June 2004....

Early Bookworks

Bern Porter: In 1920 I started making books. Up in Maine my problem was the simple act of reproduction. In those days, Xerox had not arrived and mimeograph was very crude—certainly too expensive for me—and rubber stamps had not yet come in or certainly were not available in Maine, so my problem was to draw them by hand. This meant writing, printing the text and making the illustrations. And it meant an edition of five copies, an incredible work in terms of hours and of effort.

The trick was then to condense, to re-digest, to state and to hand letter and to hand-illustrate an edition of five copies. My first such book was done in 1920. It was hand-sewn, hand-drawn, hand-lettered, and it had a slipcase which was also handmade. And since I had difficulty with titles in those days, I simply called them numbers, like 179B, and the next book, of course, was CD21. Thus, every title had at least one number and one initial.

My audience was a woman who lived down the street about four blocks, and when I completed a book, I would take it down to her, and she would give me a dozen eggs which I would take back to my mother. So I was making artists' books in those days—one book for a dozen eggs—a sort of barter system, the lady who received, my mother who received, and I who made it received no money.

I continued with this edition of five, and as far as the woman who swapped the eggs for the book, some friends from Boston would come to visit her in the summer and she persuaded them that they should give me the magnificent sum of one dollar and would take one of my books back to Boston and take it into a gallery to some of their art friends and see if there was anyone in Boston who would like to swap handmade books. My first out-of-Maine client, I've forgotten his name, was also making a kind of art book (I don't think they were as sophisticated as mine) but he was playing with words and putting them on numbers.

Those were very rich years (1920-21) in Maine, with swapping and bartering and the man in Boston who sent the books to some folks in Philadelphia, who bought some books, and the next thing you know they were tied up as far as New Mexico after about a year and a half of production. At no point were the editions more than five copies. I've been told that one of these editions that I produced in those days now sells at auction for $750, somewhat different from a dozen eggs, which in those days sold for about thirty cents. I personally have no more of those, but the master collection at the UCLA campus in Westwood does have the magazine which I hand-lettered, which runs to about three pages. So, I began at the age of nine.

Mail

BP: I also did what has later become Mail Art about that time, and was receiving postcards through the mail and would take a razor blade and cut them up into three sections and then re-paste them together, so that we had a sort of a montage, in fact, of a distorted image. Later I encountered in Copenhagen a man named Diter Rot[1] from Iceland, and he and I swapped cards which we had cut and we called those simply cut cards, to make it easy. So Mail Art was beginning in those days, and I later found that Marcel Duchamp was doing it about the same time, and Kurt Schwitters in Switzerland I later learned—they were independent of each other. But 1918-1922 was a very rich time for the beginning of artists' books, Mail Art, and what we later called posters. It was always shattering to me to sit in Maine thinking that someone else was also doing the same thing, and we got to know one another, becoming friends instead of rivals. The first artists' books went through the mail and were swapped very much as we do Mail Art now.

---

1. Diter Rot (1930-1998): German-born conceptual artist and bookmaker, associated with Fluxus and Mail Art. A.k.a. Dieter Roth, DITERROT, Dietrich Roth, Karl-Dietrich Roth

HOLD ON TO YOUR HAT

As for artists' books, five copies of thirty pages was really a Herculean task! The network was basically word-of-mouth, and the mail service was considerably different in those days than it is now (probably just as unreliable), but we were quite adaptive in those days.

Suddenly, the church that my parents attended acquired a mimeograph machine. This was a marvel of the first order, because it came all the way from New York City to northern Maine, and I happened to have a job as a janitor in the church and every week ,when no one was looking, I used to turn the crank and produce some pages on this marvelous machine! I was able, therefore, to run an edition up to ten copies. This was a very advanced technological development. Mimeographing had arrived in Northern Maine!

Judith Hoffberg: *Were you as pleased with the product?*

BP: No, and neither were my recipients. They felt that the ink impression that I made with an ordinary quill pen and a bottle of India ink was far superior to the sort of muddy gray that this marvelous machine produced.

I still feel that the machine stands in the way between me and the product. Today we are involved with computers, space satellites, and communication devices, and the artist is particularly very much in the background.

We have discovered recently that we are unable to talk with one another, and the reason is that everyone has a different definition for one and the same word. So it is necessary now to examine the word. And some of us are making desperate attempts to do this. The word is in a bad way in our culture. This is a universal problem. Few people understand one another because this difference of word meaning to each person exists, and I feel it is time for the individual word to be examined.

I find this is a product of our time, when we have nothing but words coming out of the woodwork and all of them are generally instilling fear, doubt, uncertainty, untruth ... It is almost impossible to find the truth in this great tremendous morass of words which now surrounds us.

Interviews with Bern Porter

JH: *How did your master collection get started?*

BP: My master collection has been forming at UCLA since 1942, started by Lawrence Clark Powell [the Special Collections librarian at the time]. I was working as a physicist in Berkeley on the atomic bomb, and he was interested in having California people preserved in the library. And in this category, I am reminded that Henry Miller's wife, June, used to go from bar to bar selling one single sheet of paper about six inches square on which Henry had written some words. I was able to give some of those sheets to my collection at UCLA. I look upon those as a very important form, the precursors to the handbills, the free handbills, with the exception that she sold them, and she was doing this entirely around 1924 or so.

    I personally add to the collection at UCLA every six months or so, with the result that by now it is very considerable in volume. I have a secondary collection at Colby College in Maine where I graduated and in the case that I have a certain kind of split personality ... I also have a split collection: a master collection on the West Coast and a secondary collection on the East Coast.

    In my isolation in Maine, it seems to me that I just plowed ahead in a great number of forms and sort of called them my own and sent them out on waves, so to speak, and if other people were independently doing the same thing or later imitated me or followed me in any way, I was very pleased and surprised. But in general I worked alone all these years.

PHYSICS FOR TOMORROW
Bern Porter

in architecture

in art

in communication

in literature

in music

in poetry

in sculpture

in theatre

Foreword:

    Since 1932 and early work on The Map of Physics (Central Scientific Company, Chicago, 1937) I have been endeavoring to extend the principles of physics into the humanities. Not having the benefit of academic facilities, a laboratory, or funds, I have been obligated to Wolfgang Paalen and Richard Bowman in painting, Herbert Bayer in communication, Paulo Soleri and Buckminister Fuller in architecture, Antonin Artaud in theatre and Frederick Kiesler in sculpture for their independently conducted efforts over the same years which have not only countered parallel to my own work and often unbeknown to me, but have broadened and authenticated the empirical base on which my views rest. The first section was published as The Union of Science and Art in 1947 and embraced the role of physics in drawing, etching, lithograph, silk screen, oil and water-color painting, and photography. Subsequent and continuing work has allowed release of the second section at this time covering the contribution of physics to literature, communication, music, theatre, poetry and architecture.

    Correspondence is invited.

Bern Porter, Burnie, Tasmania, 1959

SCIARCH:

Air directed under pressure in moving columns, sheets and walls define and enclose areas. Transparent, such constrictions restrict no views, passing sunlight freely in all directions. Insulated, it retains its own temperatures, excluding all others. It is snow and rain proof, repellent to insects and animals. The air may be heated for winter, cooled for summer, or colored for privacy.

Light, natural and colored, projected in columns, sheets and walls define and enclose areas. Transparent and semi-opaque, such constrictions restrict no views, allowing free passage in all directions. Ultraviolet and other radiation types provide treating, cooking, vitamin farming, restoring, healing and curing interiors.

Heat projected in columns, sheets and walls define and enclose areas. Treating effects common to light are also available.

Water dropping can make walls, columns and sheets just as water projected upwards in columns, sheets and walls enclose areas. Special purpose interiors follow.

Ethers, gases, compressed and ejected; natural, colored, mixed and seeded in air offer potential in walling and sectioning. Unusual features are feasible.

Sound vibrations, diffused and concentrated, comprise interesting psychological partitions and extra effects.

Sun provides heating, cooling, lighting, drying, washing, cooking and locomotion power; can be projected light into screens, partitions, walls and special results.

Wind can also be harnessed to produce equivalent effects.

Tide, its changing levels, provides power generation and floating pressures for interesting manufacts.

Unusual sites such as mountain slopes for sun collectors, mesas for cities, lakes for floating structures, gorges and canyons for bridge-dam-dwelling combinations can be utilized. Bridges, dams, stadiums, reservoirs, cathedrals and other massive edifices and piles offer special use, first for their intended function, then as highways, highway supports and sheltered accommodations of diversified character. Terraced dwellings are erected above highways with offices and workshops below. The anchors, piers and abutments are particularly useful for accommodations, gardens, look-out parks and view points.

the union of science and architecture
1947-1959

SCIART : SCIDRA : SCIPAI

Finite worlds of infinite reality and beauty revealed by the tools and discoveries of science are ripe for aesthetic development.

Of light, besides the commonly employed natural and artificial, there is the polarized, the radiating chemical, mineral and radioactive types along with x-ray, cosmic and nuclear-particle beams with all related electro-optical phenomena.

Of other vibrations, there are the natural, the mechanical oscillatory; resonant and supersonic sound; the entire frequency range of electrical and thermal waves.

Of movement there is mechanical and electrical accelerations to light speeds; nuclear, gravitational and magnetic interactions; the mechanics of flow and change in matter.

Of phenomena there is hysteresis, electrolysis, isotopy, relativity, entropy et cetera, et cetera; of devices there is no end: cyclotrons, stroboscopes, cynometers, spectroscopes, cloud chambers, tonometers, diffraction gratings, x-ray tubes, electron microscopes, et cetera; likewise, phases of science as meteorology, hydrostatics, crystallography, histology, aerodynamics, astro-physics, metallurgy, et cetera.

From such as these, their similar related effects, the manifold variants they suggest, stem the heretofore unrealized textures, patterns, forms, devices and techniques comprising Sciart.

It forges new reality via the depiction of these individual materia, or scientific-creative combinations of them like photography and poetry, topography and portraiture, calculus and art, et cetera, if stated by conventional methods, but, more preferably, when expressed on non-rectilinear shapes having other than plane surfaces and with media not normally employed in art.

the union of science and art
1939-1948

SCICOM : SCILIT : SCIPOE : SCIMUS

With the alphabet a crude device, unrelated to sound, vision, meaning, reading and understanding; with the world's people in need of one language, physics can determine the mechanics of hearing, listening, registering and responding in the cases of vocalization, can determine the mechanics of seeing, reading, registering and responding in the case of written communications.

Included in the determination are which vocal sounds carry all aspects of hearing, listening, registering and responded to the fullest, most direct degree, what letters convey the quickest, complete message for response, what images or symbols convey a total message, visually, audibly for the intended response and from all peoples or races.

The result will be, first, the simplified alphabet, the abbreviated word which when put into groups that fit an eye space will provide a break between jumps and an equal number of jumps per line. The second and alternate lines will be indented, and other patterns of line arrangement made to provide most direct responsive impression under all conditions of impact from page, screen and billboard.

There will follow a phonetic written language using visually efficient symbols, the symbols in turn to be a picture language or symbol and more correctly a scientifically devised design or letter and figure shape having in all probability little or any relation to the present alphabet, figure and symbol shapes…a physical or physics tongue and sign so to speak, with certain symbols to denote whole passages of meaning in much the same manner that "57" can be taken to mean a whole range of canned products from Messr. Heinz of the current vogue.

Current limitations of the voice, ear and eye will be expanded and both letter and figure symbols will provide for responses from persons moving at varying speeds.

Included in a special department of the new expression technique is a subject matter truly scientific in source spring yet treated in a dramatic and axiomatic way as the current plot of characters and life situations. Equally, literature becomes not a matter of type images, transmitted sounds or film projections but a series of (a) light signals, flashes and fusions to the eye, (b) sound signals like short dots, long as dashes, to the ear, or (c) combinations of both, the latter being conveniently instilled in periods of sleep or rest.

the union of science and communication
the union of science and literature
1949-1959

HOLD ON TO YOUR HAT

SCIMUS : SCICOM : SCILIT : SCIPOE : SCITHE

Scientific subject matter, techniques, materials, procedures and advances extend the field of music.

Subjects are cited under Sciart and themes from the progress of science abound.

Techniques utilized after making modifications common to Scicom, Scilit, Scipoe alter music radically while extensions of its vibration both below and above the present hearing spectrum with heretofore unrealized devices and tone scales, many arising from the physics of electronics and its manifold advances, further change its nature and content.

Materials and devices follow from the perfection of new substances having tonal qualities; the development of tone-mixing, tone-regulating devices; the mixing of sound tones with colored gases, lights and water; the automatic production and control of heretofore unheard sounds, imitations and faithful reproductions of existing sounds; the projection of both types of sounds in directional beams, sheets and bulk in combination with heat, light, liquids and gases.

Procedures and advances stem from the scientific themes and methods developed and extend as experimentation in materials, construction, electronics and engineering proceed with studies on the mechanics of hearing supplementing the progress.

the union of science and music
1947-1959

SCIPOE : SCILIT : SCICOM : SCIMUS

Physics projects poetry beyond the typographical entrapment traditionally circumscribing it as a visually read experience.

Physics provides a method, principle, terminology, spirit and advance to be incorporated into poetics for enrichment and extension.

Enhancement is further possible with electronic recording in the pure word state by microphone, tape, record and loudspeaker for transmission to the ear.

There is extra word embellishment by music, vocal and instrumental, largely unfortunate, extraneous, or irrelevant, which should better be replaced by natural sound accompaniment in keeping with the words or mood.

There is projection of words by films and slides, with or without mood or moving illustrations, to the accompaniment of sound, producing an eye-ear reception.

There is the reduction of reading or seen word to symbols; reduction of the spoken or heard words to symbols; the fusions of both being received as flashed images to the eye and sound signals to the ear, with combinations creating a wholly sensory reaction exceeding in depth of response that found in music, hence Scipoe. There is the wholly intravenous thought projection and reception between the poet and his audience, hence Scipoe extended. There is the combination of poetry with photography (photo-poems), with other picture types (pic-poems, drawing-poems, art-poems), with games (game-poems), et cetera.

the union of science and poetry
the union of science and literature
1940-1956

SCISCU : SCIART

Scientific subject matter, techniques, materials, procedures and advances extend the field of sculpture.

Subjects are cited under Sciart.

Techniques beyond conventional moulding and casting include brazing, welding, cementing, fusing, evacuating, polarizing, magnetizing, electrifying, anodizing, annealing, etching, electroplating and other processes common to physics.

Materials beyond conventional plaster, wood and metal include plastic, resin, gum, foil, masonite, bakelite, plasterboard, crystals, gems, minerals, ores and other substances common to chemistry.

Procedures and advances current in physics, chemistry, geology, biology and engineering can be scrutinized monthly for adaptations to sculpture.

Bringing sculpture up onto walls; out into and in contact with the environment of onlookers; making it also an object to be used; for easing hypertension, for sound and musical purposes, as screens, partitions and walls, a piece of furniture and combination thereof, a fireplace, the cooling and heating system, altering its concept to sociological conditions include the extended uses.

the union of science and sculpture
1946-1959

SCITHE : SCICOM : SCILIT : SCIPOE : SCIMUS : SCIART

New theatre follows from the use of language, literature, poetry, music and art admixed with the principles of physics as described for Scicom, Scilit, Scipoe, Scimus and Sciart.

Scicom, Scilit, Scipoe add: vocal, instrumental and choreographic scripts; media for projection and communication.

Sciarch adds: open-air, seat pits, conical in shape with aerial center stage; open-air circular set seats with overhead rotating beam stage; square enclosures with performing balconies on the four inside walls; cylindrical enclosures with an inside spiral performing stage; weather control for open-air performances; sound, heat, light, gas, liquid, key-operated organs.

Sciart adds: subjects for play, opera and dance development; motif for masks, costumes, drops, properties, programs, billboards; a new dimension and atmosphere.

the union of science and theatre
1939-1959

BERN PORTER 1959

Introduction to Xerolage 16
Bern Porter

One is a plasma of energy. Simultaneously living/dying, oscillating/passive, expanding/contracting, sending/receiving, alert/dead, seeing/blind, hearing/deaf, breathing/exhaling.

A four dimensional space form of height, length, breadth, time.

Proceeding, returning with such speed in an orbital arc of such magnitude that the path is a straight line.

And at such a speed there is no past, present or future.

Only a NOW of/for fullfillness, all or nothing.

The plasma is/was for ever present, here from the beginning of the beginning.

Created, set up by natural formulation and design.

It will always be, never die, cease, extinguish or reduce in strength.

Its characteristics were instantly and

permanently formed at one's inception, never to
be altered in any way by parents, schools, churches,
experiences or events.

    It is

    Nine to ten months later it was given your name

    It is you, yours

    And all the knowledge, experience you have, had.
Will experience.

    Its container is your physical body which at
death will drop away while your plasma will continue in
orbit through space at speeds of no present no
future.

    You will always be.

Bern Porter
Madison, Wisconsin 3/21/90

Significant Content
Bern Porter

If pure feeling born of a relaxed state is the source-spring of creation, what is the nature of content realized?

First, the fountainhead of inspiration accepts no compromise: its will is always done. While the reasons for this may seldom be evident, the resulting action is for the best.

Second, works so motivated contain mainly good, purposeful elements. Corrupt exploiters and interpreters plotting otherwise fail to change this.

Third, inspired works provide the greatest number of people with maximum benefits. Content thus becomes more than significant: it is a part of everyone interested.

Fourth, executors of the revelation are both its conveyers and guardians. Highly privileged, they are not accumulating personal gain, but are the means to an end–intermediaries of the spirit and the people. The devil take these artisans if they distort the inspired image or misuse their talents in developing it.

Examples of creative endeavor may clarify these points.

War is more degrading than noble. It reveals bestial men: how inconsequential and meager are their achievements. This important theme, needing amplification and publicity, has been variously assigned to Beethoven, Sibelius and Shostakovich. That they fulfilled their charge for many peoples and eras in spite of critical interpretations and piracy is well known.

Simple things like stars, trees and sunflowers were left to Vincent Van Gogh. This trust he natured as one obsessed and transmits today in spite of contracts, copyrights and agreements of conniving dealers.

Circumstances attending the development of light bulbs, air coolers and automobiles follow similar patterns. The cartels, patents, tariffs and other paper codes man devises for profit taking–and suicide–do not wholly mask the significant content of the most prosaic articles. Good intent radiates from them and for just reasons.

Curiously the inspiration for a "G-Block Bomb" is not granted anyone unless such negation has been ordained. When completed, its potentialities for constructive use outweigh the pernicious elements.

Thus the core of created form is a pre-willed, purposeful content that is self-perpetrating as long as anyone will approximate the awareness of its creators.

Bern Porter

The Preposition Song

(sung to the tune of Yankee Doodle)

WITH ON FOR AFTER
AT BY IN
AGAINST INSTEAD OF
NEAR BETWEEN
THROUGH OVER UP ACCORDING TO
AROUND AMONG BEYOND INTO

STILL WITHIN WITHOUT UPON
FROM ABOVE ACROSS ALONG
TOWARD BEFORE BESIDE BELOW

BENEATH AND DURING UNDER

BY MS. TESSIE

See(MAN)TIC / Bern Porter
Amy Hufnagel

January 15, 1994 – March 20, 1994
Robert B. Menschel Photography Gallery (issue #35)
Produced by Light Work, Syracuse, New York

This summer, during a video-taped walking tour/performance in Belfast, Main, Bern Porter entered the local bank at lunch hour. He stood in line. He waited and waited and waited and finally it was his turn to approach the teller. He approached her, smiled, and said hello. Then he turned and left the bank.

If you walked past Bern Porter on the street, in his coastal hometown, dressed in his "Poet Laureate" brown robe carrying a huge cardboard "A," creating and reading his poetry on the street's corner or leading his own parade, you would probably stop to contemplate whether he was a genius or a crackpot. If you knew Porter, you'd know that he transcended even these most conspicuous labels. His actions take the extraordinary and rebellious, creating a hummm of confusion about the artists' personae in our culture. Entering the bank to say hello alters the behavioral and cultural norms we expect from people in this space and this act is synonymous with most of Porter's art and life. Both are unorthodox. As one who invents and harvests eccentric and bizarre ideas that make sense, Porter has spent seventy-five years making art from objects, images, and situations that most would rarely contemplate let alone have the confidence to execute. Porter is a bundle of mad momentum, an atom hurled by the intensity of his own energy, making artist books from newspapers to trade for milk and eggs in his hometown in Northern Maine as early as 1916. Now Porter is the published author of sixty artist books and poetry collections. He has also published the work of many emerging authors, performed and shown his work extensively, and he is the subject of many authors' writings. See(MAN)TIC is an exhibition that concentrates on Porter's photographs from 1937 to pieces made specifically for this catalogue and exhibition. The culmination suggests a lifelong career dedicated to producing art and ideas in a wide variety of disciplines that demonstrates he "cannot escape the feverish inventions of the mind."

HOLD ON TO YOUR HAT

Porter is an artist and physicist whose work requires that we "show how big our brains are." His acute imagination, and the desire to escape his feelings of isolation which result from being so forward thinking, have propelled him to seek those in the avant-garde who "cut the crusts off accepted forms" ranging from the Surrealists to the Dadaists, the Beats to the Fluxists, the Abstract Expressionists to the (Post) Modernists. As a Sciartist (scientist/artist), he is constantly experimenting to find the potential behind, or the possibilities within, the "matter" which he encounters. His art is the result of his constant experimentation. His biographer James Schevill writes, Porter is always conducting Research, "examining the arteries of American power and culture where they flow visibly. The images he witnesses are reflected inevitably in his artwork."

One of the central locations for his current research is the trash-can of the Belfast, Maine post office. An avid mail artist, Porter visits this "lab" daily. His base data is rich from the abundance of junk mail findings; once cut and altered, they form the base of his collage work. Perhaps from his physics training or equally pertinent social belief that conformity and specialization should be avoided, Porter asserts that something new has to be produced in order to alter the old matter. There is no denying that the material in the trash could stand some improvement and Porter finds inspiration in reclaiming trash. His "Yankee/Puritanical" Upbringing may be the root of some of his trash ingenuity, but his personal experience as a physicist working on the Atom Bomb as part of the Manhattan Project in the early 1940s also contributes. The explosion of the bomb in Japan in 1945 "blasted a hole through his idealism" as he realized exactly what he had been building. He resigned from the Manhattan Project the next day disparaged, disillusioned, and devastated by his personal role in the destruction of humanity. Porter vowed, from that day on, he would contribute to and reclaim humanity rather than aid in its destruction. He says, "I merely felt that I could and should do more good. The reaction from destruction was simply to do something constructive with what limited funds and talents I had."

Because Porter is physicist, his knowledge of the elements, formulas, and technology used to explain the universe are essential to understanding his connection to photography. While spacial relativity, light composition, and lens optics are central to his process, Porter also explores the most esoteric ideas of order and chaos. His artistic expression seems to shift back and forth between making order out of chaos and chaos out of order, always relaxing or intensifying our expected perceptions. Take for instance the *Manhattan Telephone Book*. This book is the epitome of order and referential information. Introduce the energy of Porter, and the book is cut, clipped and shuffled upon a tabletop, adhered into a montage of impromptu found visual poems and word pictures. Porter uses the same process with travel brochures, magazines, junk mail, newspapers, and objects to make everything from artist books like *Sweet End* to photograms and bottle poems. The series of eight "Bottle Poems" are medicinal bottles filled with found material like pretzels and nails, photographs and shredded paper, bird food and fabric, offering visually "the cures for what ails ya." These poems suggest that the contained chaos is the antidote for our instant gratification society; we are, after all, a throw-away culture creating chaos in the trash can so that we appear ordered. Porter, however, "feels the semblance in chaotic form" and notes his distaste for ordered reality. In turn, his Bottle Poems offer us a revised dose of our own medicine.

Porter understands that what holds things together is usually not so different from what breaks things. His desire to tear down the gallery walls, discard the frames, dig through the trash, break words into sounds and pictures, transform the audience, build alternative environments, teach innovation, and issue proclamations all relate to a single concept: he can make something new by breaking up and rebuilding that which exists. Using founds, Porter invents new materials and alters structures by changing the environments and relationship of the combined elements. As his friend Buckminster Fuller said, "Our world is held together by tension" and its structures can be altered and still support the weight of humanity's demands. If art, or the physical world, cannot support and exist on unique and ever changing forms then it will, in Porter's words, "collapse to zero."

In his 1948 Sciart Manifesto titled "The Union of Science and Art," Porter begins, "Finite worlds of infinite realities and beauty revealed by the tools and discoveries of science are ripe for aesthetic development." He claimed that such combinations as photography and painting, topography and portraiture, photography and poetry, and biology and sculptures create yet unrealized textures, patterns, forms, devices, and techniques. To further this idea of the infinite realities available to our cultures Porter proclaims, in his autobiographical manifesto *I've Left* begun in 1954, that art has to be everywhere, as is the military and technology, in order to escape the false traps of specialization, structure, bureaucracy and stagnation. According to Porter, this art can be created from the way one gets dressed, by printing art and poetry on playing cards and stamps, and by actively creating new languages, to name a few.

The rational theory for Sciart and his art's suggestions for social reclamation do not cancel out the serendipitous qualities in Porter's expression. The descriptions of his artistic process show a playfulness in the subconscious, an approach often associated with Surrealism. In 1935 Porter met Salvador Dali and Max Ernst in New York City, igniting his interests in non-objective and Surrealist modes of artistic experimentation. In 1937 Porter made his first photographic work of double and multiple exposure painting-like montages, like My Love For You, beginning his fascination with image manipulation rooted in science but also in a forced lack of rational control. But as the Tao of Physics proclaims, "It is never nothing, but always something."

Amy Hufnagel
Assistant Director
Light Work

(All quotes above are either the words of Bern Porter, or they are excerpted from the biographical writings on Porter by James Schevill. Biographical formula by Amy Hufnagel.)

Books by and about Bern Porter
Andrew Russ

James Schevill. *where to go, what to do, when you are Bern Porter: A Personal Interview.* (Tilbury House Publishers, 1992) 326 pp. $16.95

Bern Porter. *Sounds that Arouse Me.* (Tilbury House Publishers, 1993) 161 pp. $9.95

Bern Porter. *Less Than Overweight.* (Plaster Cramp Press, 1992) 500 pp. $28.00 (POB 5975 Chicago IL 60680 / press run: twenty-eight copies)

    Anyone who has an interest in reconciling or integrating the practice of science and art will want to know of the pioneering work of Bern Porter. Born 1911 in Maine, he trained to be a physicist, eventually getting a Masters of Science from Brown. He simultaneously absorbed what he could of the arts and continued educating himself as he worked on cathode ray tube technology in his pre-war job. During World War II he had a role in pioneering uranium separation for the atomic bomb, and became the first to publish Henry Miller in the U.S., with an anti-war tract entitled *Murder the Murderers*. This was followed with the publication of a work by Kenneth Patchen. When Porter saw the full effect of his research efforts in the explosion of the atomic bombs in Hiroshima and Nagasaki, he left professional science with a guilty conscience and set up the Contemporary Art Gallery in Sausalito, California. By now Porter's lifelong pattern of self-sacrifice for the promotion and publication of art he believed in was firmly established. Porter had also started producing his own body of work: poetry, experimental essays, surrealistic photographs, collages, found sculptures, architectural sketches, and the found poetry that he is best known for. In 1950 he published the sciart manifesto *Physics for Tomorrow*, which is his plan for the integration of science into art, essentially the finding of applications for scientific phenomena in the production of beauty.

    Where to go, what to do, when you are Bern Porter chronicles Porter's life in its cycles of greater and lesser financial poverty as he found ways to produce and promote creative endeavors that mattered to him. His eventual production included an anti-utopian manifesto entitled *I've Left*, forays into performance art, sound poetry, mail art, and video. In addition to the ongoing production of literature and visual art, Porter also produced a mammoth regional report for Knox County, Maine. He worked for the Saturn V project[1] for a short time and founded the Institute of Advanced Thinking, a network of scholars from the various arts and whose membership was closed to members of academic or governmental institutions by Porter's preference. The catalog of work is really too long to summarize here, and Schevill's book has a lengthy bibliography and a generous supply of pictures, both of Porter and of the works he has created and promoted. The last chapter is devoted to an essay on Porter's *founds*, or found poetry, including discussions of Porter's radical readings of them.

Much of the book describes the great obstacles Porter faced in promoting his art: varying degrees of poverty, occasional surveillance by the FBI (interesting excerpts of his files are included), the conviction of his father for a sex crime (leading to Porter's period of self-imposed exile on the remote island of Guam), and unsympathetic employers and neighbors (some of whom are quoted in the FBI files). The moral is that one must pursue one's vision in spite of such obstacles.

Sounds That Arouse Me is an extensive sampling of Porter's own writing, excepting the visual—only one found poem appears at the end. This sampling includes finished poems, parts of books, manifestoes, excerpts of interviews, and a couple letters. The writing ranges from the dumb (e.g. "Blank Verse": ∧∧∧∧∧∧∧ / ∧∧∧∧∧∧∧∧∧∧∧∧ / ∧∧∧∧∧∧? etc), to the incompletely thought out (e.g. "Statement"), to the moving and significant (e.g. "Why Don't You Use the Trail?", "Sciart Manifesto", "Me"). The personal reminiscences of his encounters with people like Henry Miller, Albert Einstein, J. Robert Oppenheimer, and Gertrude Stein are also very interesting. The biggest shortcoming in this anthology, ultimately, may be the neglect of the found poems, of which we have but one or two examples (if you count the rather arresting "Found Story"). I take issue as well with the resetting of *What Henry Miller Said and Why It Is Important*, as it leaves out the all-important white space in the original. And given the critical role of the visual in Porter's poetry, I find myself wondering what

---

1. Saturn V was the model of NASA rocket that carried both the Apollo and Skylab mentions into space. Thirteen were launched between 1967 and 1973.

else was lost in the process of putting all of his work into the same typeface. While this selection is interesting, many of the best works are quoted in detail, if not in full, in Schevill's biography, making that the better book to read first.

Less Than Overweight is the latest volume of founds produced by Porter. It happens not to be part of the ongoing series of seven volumes of founds—starting with 1972's *Found Poetry* (Something Else Press) and *The Wastemaker 1926-1961* (Abyss Publications)—that Porter outlined in *where to go, what to do, when you are Bern Porter*. However, the book's scope and length (five-hundred or so unnumbered pages) is about the same as for the other books in the series. It is Porter's stated preference that to properly read this and other such books, one should dive in, find a page at random, read from it, dive in somewhere else, read from there, and in this manner leap around, connecting various pieces as one wishes. There is a certain ergodic randomness in the arrangement of pages that renders linear order irrelevant.

The early founds were abstracted from ads and magazines and newspapers, with the words edited into absurd bits or with pictures and captions juxtaposed in odd ways. There is still some of that element in the present volume, but here we find a greater visual element, as well as more of a willingness to reproduce some items just as they are. There are full-page ads from a local hardware store, a letter explaining that "Friend Porter" is eligible for prizes of "up to $40,711.00," children's drawings, etc. If it seems like a collection of random images and pieces of paper that Porter has gathered, indeed it is that. You can read it as a kind of archaeological project and see elements of Porter's life in these fragments. In fact, the autobiographical aspect in this book is much stronger than in the previous books I have read (*Found Poems* and *Here Comes Everybody's Don't Book*). There are several sheets with small founds sent off for various mail-art exhibits, all with the added notation—"founds by Bern Porter, return after use or non-use;" one found is recycled from its previous context as a *Score* broadside; as well there are a few notices of Porter reading and a couple one-page bios. There are also some scraps with Porter's handwriting—some sums and numbers—and several papers from correspondence and court records involving a suit between Porter and one Raymond Morrison. One could dive around looking for details of aspects of the dispute, and begin constructing a story (the arbitrariness of the ordering of pages means one has to sort out the chronology), giving one element of a running plot in the book, but not one that gets to any real resolution. If anything, it is a theme of legal action and money, and you see the flux of that theme as it runs through that book.

There are several such flux lines, or tropes: lotteries, children's art, architecture, chromatography, a sewing pattern, hardware, Mail Art, clothing, threshers. Often the flux lines intersect, such as a child's drawing of a "helping hand" followed several pages later by a "wet paint" sign with the outline of a hand; or a page with an appeal for funds from Amnesty International accompanied by a "Grand Prize Winner's Authorization." The book's pages are carefully disordered so as to avoid a logical progression, leaving it to the reader to search through and make his/her own connections. The reader is even given clues to do this, by way of, for example, an image cut in two and reproduced on separate pages, or an image that appears twice, in differing contexts each time. The overarching theme, if any, is commerce, though there is material that clearly doesn't fit the theme. But this is fitting, as Porter was largely an observer outside the American economic circus rather than a performer in it. This book demands a phenomenological method of reading, where the reader has to jump in at some point, see what's there, and move through the book, gathering information until a sense of the intelligence behind the selection emerges.

Seth Tisue[2] has lovingly produced this book, including color copies of Porter collages at the beginning and end, binding the book as a hardbound edition, adding a laminated plastic dust jacket with a color front cover illustration, a page of cut-out phrases from hand-written notes by Porter about producing the book, and an excerpt from *I've Left* to illuminate the text. This was the final publication of Tisue's Plaster Cramp Press, and is limited to twenty-eight copies, distributed to those who are truly interested in Porter's work.

---

2. Seth Tisue (b. 1971): Chicago-based publisher, writer, and software engineer. Editor of Plaster Cramp Press.

Review of Bern Porter's I've Left
Andrew Russ

Bern Porter. *I've Left: a manifesto and a testament of science and art.* Introduction by Dick Higgins. (NY: Something Else Press, 1971)

Written before C.P. Snow's "The Two Cultures" lecture,[1] *I've Left* (1954-59) is an early attempt to unify science and art by a man who spent his career with one foot in each culture. A good example from Porter's resume is his publishing of Henry Miller's anti-war tract Murder the Murderers in 1944 while working on the Manhattan Project in Oak Ridge, Tennessee. There is an impressive list of names dropped on the inside back cover of *Xerolage 16* (Xexoxial, 1990) that can give you an impression of the broad range of Porter's publishing interests.

Bern embarked on *I've Left* some ten years after his Oak Ridge days. Having given up serious physics work, he opened an art gallery in Sausalito, traveled to Guam and lived there for three years, and toured Japan, especially Hiroshima in order to see the effects of the bomb that he had worked on a decade before.

His interest in the arts was, as is the case with many physicists, bent towards the avant-garde. This is partly because physicists like to look at the extreme cases of their theories—that's where the equations can often be solved, and that's where you can see if a theory breaks down—falsifiability is a pretty big seller in the physics community. The avant-garde projects itself onto the frontier of ideas, just as ambitious physicists do.

I've Left is an exposition of Porter's ideas on the unification of Science and Art, which he terms Sciart. This means both putting technological developments into creating new forms of art and putting aesthetic and scientific ideas into practice in daily life.

The official starting point of Porter's unification is his SCIART manifesto, "Physics for Tomorrow," variously dated from 1939-1948, revised in Japan in 1954,

---

1. Charles Percy (Baron) Snow (1905-1980): English novelist and scientist whose *The Two Cultures and the Scientific Revolution* (1959) posited that the communicative divide between the sciences and the humanities was a serious obstacle for social progress.

and published in Tasmania in 1959, and finally republished as "Physics Today" by Roger Jackson in 1999. It begins: "Finite worlds of infinite reality and beauty revealed by the tools and discoveries of Science are ripe for aesthetic development." This proclamation is followed by a list of hypothetical discoveries (though not an exhaustive one) that Porter considers aesthetically useful. But his approach in *I've Left* starts with a mystical vision:

> I finger zero, readjust my couch in a void that sloth built, the better to do nothing ... Obsolescence revolts me. The alleged modern is a repetition of the ancient decorated in chrome, styled with air-flow and color-engineered to abomination ... Thus, communication-wise I junk drum beats, smoke signals, semaphores, flag codes, light flashes, telegraphs, telephones, radios, television sets and all other such systems, devices and developments for my own sensory organs wherein desiring to make known my wishes I merely think them in a frequency universal and in a tongue world known and whomever wishes to hear, receive and understand does so. The spoken word, printed and tele-dramatized word becomes a particle of thought energy. The drawn, photographed, painted and kinescope-picture becomes more of the same. All of the devices of locomotion, subterranean, surface and aerial equally reduce. I am at all places, in all forms, at all times. What were books became word sequences screen projected, then free-floating vibrations which impinged upon my mind as I desired them. What was art left museum walls to become gaseous fusions in color similarly projected, then all prevailing rhythms of radiant energy that stimulated my eye whenever I wished them. What was poetry became equally transformed to responses for feeling. Architecture became constructions of ether and light. Clothing a logical extension of skin without embellishment. Theatre a pageant of masked spectators. Automobiles, body rockets. Toys, fondling in the dark. No civilized thing was left unmodified or unreverted to its natural, logical and true state. I transformed the world and in so doing I found myself (pp. 1-2).

Interviews with Bern Porter

This utopian vision has eight chapters, each different, each corresponding to a thing transformed: poetry (and literature), clothing, theatre, architecture, visual arts, food, stress, and the automobile.

The first chapter covers poetry, but does so in an odd way. Porter writes a first-person narrative of a man who prints poems on fine press postcards and books and finds he can't sell any. He starts to print poems on playing cards and chessboards, but their users complain. The narrator presses on, printing poster poems and photo-poems fusing word and image. "Art-Poems likewise came into being. Drawing-Poems. Chem-Poems. Math-Poems. Any form that existed; those that didn't." He goes on—enlarging cigar-wrappers and theatre tickets for the purpose of printing poetry on them, also laundry lists, match covers, gum wrappers, and toilet paper. "No wasted spaces." He progresses into a delirium of poetry saturation-bombing:

> I spewed poetry without words, without sounds, without punctuation, without meaning. Poetry died. In its place arose dots and dashes in full color, cinescope projected with mail-slot screens; intermittent spurts of chemically-dyed ether, alpha radiated Geiger recordings emanated from my diaphragm with every third vowel crystallized. Textual poetry was at last off the typed ms, off the printed page and jet soaring ... The lights, the dots, the dashes, the equivalents of words became meanings, tones of meaning, and the image of meaning. Poetry was at last intrinsically itself full born but umbilically tied to its source spring ... Poetry was now in the realm of physics and I termed it SCIPOE (pp. 4-6).

Perhaps some will read this as an imperialistic attempt to effect a subsumption of poetry into physics, but once the implications of SCIPOE are realized, poets become well-paid state employees who create endless surpluses of words, which the narrator then dissolves. The poet studies the typical motion of eyes and patterns line lengths after them, printing in color to free the printed page of its "black and white dullness." And lest we think this is merely some fantastic raving, there are footnotes citing some of Porter's other publications. But the ecstasy of communication and meta-communication continues, culminating as follows:

"I pressed the membrane separating man and God closer and closer to the Ultimate and indeed felt occasions when I had pierced the barrier sufficiently to more than look past. The first to raise himself by his own boot-straps I was thus able to see, hear, taste, feel, and smell all things at all times. Through the combined use of these facilities I became all things at any time in any place. Thus endowed, being so, I became me" (p. 11).

In short, Porter envisions a kind of transcendence through technology. The aim here is not to write any particular kinds of poems or be bound by any particular poetics, but rather to fully explore the potential of the medium. This exploration is the ultimate goal. The process is the product.

The second chapter concerns clothes. Here Porter shifts from the transcendent to the every day, but in an idiosyncratically Porter way. He begins by cataloging all the motions he makes in the course of getting dressed:

"I pull on my trousers, draw on, around and tuck down a dress shirt, making a total of four layers of cloth about my waist with only one layer on my arms and legs. The adjustment of trousers' front and belt after getting them on consumes in all eight separate motions . . . At this point I have seven pockets. I have used thirteen buttons. I have gone through forty-one separate motions" (p. 12).

Porter continues this Tayloristic effort to keep track of the motions he makes and the buttons he buttons while putting on the eleven separate garments necessary to be properly dressed. His improvement to this situation is to combine pieces of clothing until he comes to consider fabricating a "one-piece, all-occasion suit" with "no cuffs, collar, tie, belt, double thickness at any point, and only three pockets." He further envisions engineering the wear-prone spots to prolong their durability, and even further, he foresees that "human skin can never be improved upon, might under the right conditions be made to grow to suit better a given purpose" (pp. 14-15). Porter analyzes the problem (in this case what to wear) in basic terms and proposes some ideas for solutions (which may or may not be so pragmatic).

Porter next goes about revolutionizing theatre, creating SCITHE. He goes about this by first throwing out everything that comprises theatre:

"By everything I mean the auditorium, the stage, the settings, the orchestra, the actors, the costumes, the dialogue, the lighting, the sounds, the ushers, the balconies, the tickets, the box office, the directors, the managers . . . the entire everything in total, all there ever was . . . Thus completely liberated I was free to construct the only plausible theatre, the one based on physics and called by me scithe (p. 16)."

So what is the new theatre? It has 1,200 seats arranged in a large spiral. The seats are swivel seats that can rotate 360 degrees, and these seats are heated and lit with temperatures and colors determined from "an unseen organ." The lights and heat and additional sound tones would generate emotional sensations (brightness, warmth, anger, soothing, fright).

"Moreover the heat and light could be spread continuously, intermittently; spread in waves, sheets, in torrents, in arrow-like advances, directed, piercing, streaking under, over and about the audience at the will of the organist playing from a composed score for each presentation" (p. 17).

Each member of the audience (or feeder) would bring a mask and costume of his/her own design and choice. The feeders could chant or sing or groan or grunt or cry or talk.

"Obviously this theatre of the spirit, for it was that and more, induced trances, orgiastic releases, mental flushings, emotional washing and other highly desirable internal cleansings which would clear men of hate, lust, greed and war if properly conducted throughout the world" (p. 18).

Architecture undergoes the same transformation as the theatre:

"A whole series of constructed forms adaptable to work and living resulted because in my concentrated effort to reform and redirect the development of architecture I, (1) burned all existing journals, texts and monographs on the subject, (2) shot all living architects, and (3) combined architecture with physics" (p. 20).

Buildings in a city block are: "compacted into a single round pyramiding structure ... with exterior, spiraling-walkup passages through sufficient floors to contain all the inhabitants of the prior formation. Residual areas were profusely landscaped into parks" (p. 20).

HOLD ON TO YOUR HAT

Porter envisions moving sidewalks, fans and ventilators to force "smogs and fumes skyward," windowless buildings with climate control, building surfaces designed to collect solar energy (both for electricity and heat), and rooms without corners—"shell type dwellings readapted from eggs, the shapes and habitats of crustaceans, insects and worms" (ibid). He then proposes moving into underground buildings.

Porter continues by writing that, "every house should be a self-contained, living organism free of all adjacent organisms and living directly off the environment" (p. 22). This seems to be a call to create some kind of living plant you could dwell in.[2] There follow descriptions of beds, furniture, decoration, wires (or the lack of them), and gardens. He ends by describing a church made of light and envisioning that his "techniques will save mankind" (p. 26).

Art is given a similar treatment. Porter begins with the statement, "Art has been neglected: nothing has happened in the field for centuries" (p. 26). He includes painting, sculpture, and music. He gives orders: "Get the subject off the canvas out into the air," I said. "Reverse perspective." "Discard frames." "Let the matter crawl around on the wall." "Kick in the wall and let the stuff stand alone" (p. 26).

This is a prelude for a merging of art and science:

"In these processes the standard subject matter from nature, from life, from a fused imagination of both gave way to interpretations from science as the constructed forms themselves became more scientific and mathematical. The entire advance of science to that time offered unlimited and unexplored vistas to beauty ... Working with light, metallurgy, hydrostatics and art I founded SCIART, the future of art. Experimenting with sound, quantum mechanics, electronics and music, I founded SCIMUS the future of music" (p. 28).

And this process erupts like the others above:

"Near dementia engulfed me, exhaustion and debility ensued with intermittent periods of lightning and high intensity creation in which it was impossible to even make notes of all the new heretofore unrealized developments for art" (p. 28).

The new art changes the social order by making art profitable for the artists.

---

2. See Paul Laffoley's solution to the housing shortage in "The Phenomenology of Revelation" in *Disinformation: The Interviews*. Ed. Richard Metzger (The Disinformation Company, 2002).

Three million artists are employed to design postage stamps such that no two are alike, and the artists are paid with the royalties from the stamp sales. The same goes for wallpaper, textiles, cards, calendars, gift wrappers, ads, etc. Anything with any artistic component has a royalty attached to it, and union regulations ensure that art is placed anywhere it can be used. "Artists are rolling in dough" (p. 29).

And there ensues a new era of luxury and plenty, which is followed, in the last paragraph, by a saturation where creativity ceases and the artists walk out. The resulting international crisis is solved by putting artists to work as politicians and statesmen. "The artists now fed and clothed, saved the world" (ibid).

Porter next tackles food. After considering and abandoning futuristic considerations of liquefying and injecting food at infrequent intervals, he decides to overhaul the food distribution system. Food is meant to be eaten with a minimum of preparation (i.e. raw rather than cooked, in most cases), and Porter sets up a system where people pick up foods at or near the farm or orchard. Food becomes socialized through a free market process: small producers are swallowed up by giant conglomerates that are publicly owned and dedicated to providing fresh food at near cost.

Hypertension and stress are relieved by creating Do-Das, "toys for adults" (p. 33). These can take the form of a lump of colored clay or foam rubber, designed to be rubbed, touched, and kneaded.

In the final section, Porter disposes of the automobile, which he views as the most destructive force in modern culture. (Porter never owned a car and only rarely traveled in one.) He catalogs his invective, using liberal numbers of numbers:

> "Is it not obvious that as thousands of more cars are produced every minute the bulk of congestion, traffic and parking problems could be eased by cars of smaller not larger dimensions? ... that less death and less injury from automobiles can come from lighter weights, slower speeds, less power? ... that it is superfluous to build cars for cruising when there is no place to cruise? ... even sillier to build them to go 110 miles per hour when there are no stretches left in which to attain these speeds for more than a minute at a time ... that engine powers over

80 horses are unnecessary for all but trucks ... that weights over 750 pounds are also unnecessary ... that engines should consume one gallon per every fifty miles instead of the present 9 to 18 miles per gallon ... that cars made now for attention free life of 4 to 9 months could be made to last a lifetime ... that there is no reason why the inside of a car should be an exact replica of the living room ... that a car brings out all the selfish, possessive, proud, fear and antisocial aspects of their owners ... that with no place to go at 110 miles per hour and no place to park the car upon arrival makes the automobile really just another bauble fast approaching a point of even less usefulness, excepting its major role in nose and throat ailments via air pollution, its use or effect upon cleaning conditions and dirt-free living to say nothing of its use as an end to voluntary suicide" (pp. 34-35).

Porter advocates using governors to prevent vehicles traveling faster than thirty-five miles per hour. He advocates sonar or radar devices to make collisions impossible. He advocates crash-resistant body parts and bumpers. If we must go fast, why not use electricity-driven chairs or atom-powered roller skates? It is worth bearing in mind that here Porter is criticizing the 1950's Detroit car—the gas-guzzling, tail-finned monsters cruising the roads at the time. This was before Ralph Nader, during the infancy of the interstate highway system. However, I don't believe Porter's views have changed much in the interim. This chapter and chapter two (on clothes) are reproduced in his selected writings—*Sounds That Arouse Me* (Tilbury House, 1992).

In a series of appendices, Porter gives brief, manifesto-like characterizations of his various SCI-arts: SCIARCH, SCIART, SCIDRA, SCIPAI, SCICOM, SCILIT, SCIPOE, SCIMUS, SCITHE, and SCISCU. This appendix is an extended version of Physics for Tomorrow. The following paragraph gives the basic flavor of these appendices:

"With the alphabet a crude device, unrelated to sound, vision, meaning, reading and understanding; with the world's people in need of one language, physics can determine the mechanics of hearing, listening, registering and responding in the cases of vocalization, can determine the mechanics of seeing, reading, registering and responding in the case of written communication" (p. 43).

So what should be made of all this? *I've Left* is an interesting and inspired, though often fantastic, vision of the future. It is one that embraces technology in some places, yet embraces sustainability in others.

Porter's mode here is that of the visionary. He is looking for the extent of possibilities in whichever medium he is considering at the time, not (easily) bound by the technological limitations and social conventions of the 1950s, or even the 1960s.

This is characteristic of the mindset of the physicist, the temptation to take an idea or object and run it to observe its limits. So much in physics is the testing of the limits (low frequency, high speed, etc.) of an equation or theory. In this way, Porter's basic message is that there is a lot more that can be done with, say, poetry, music, or art. There is no reason not to make advances.

Porter's other message is that technology must serve actual needs and do so without causing waste. His own lifestyle was very frugal, and much of what he writes about clothing and food and cars is a reflection of this. While Porter has a disturbing tendency to act as the dictator of his own utopia (don't forget all the architects he shot in Chapter four), his goal is to show us the greater possibilities of art and to reorder our priorities in using technology.

Unfortunately, this book has been out of print since Dick Higgins's Something Else Press ceased operations in the 1970s. The recent biography (*what to do, where to go, when you are Bern Porter*, by James Schevill) and collected writings (*Sounds That Arouse Me*) are both published by Tilbury House Publishers, 132 Water Street, Gardner, ME 04345.

Hat-isms

from *http://www.villagehatshop.com/hat-isms.html*

Talking Through Your Hat

To talk nonsense or to lie. c1885. [In an interview in *The World* entitled "How About White Shirts," a reporter asked a New York streetcar conductor what he thought about efforts to get the conductors to wear white shirts like their counterparts in Chicago. "Dey're talkin' tru deir hats" he was quoted as replying.]

Eating Your Hat

There is no such thing as a sure thing, but that's where this expression comes from. If you tell someone you'll eat your hat if they do something, make sure you're not wearing your best hat–just in case. [The expression goes back at least to the reign of Charles II of Great Britain and had something to do with the amorous proclivities of 'ol Charlie. Apparently they named a goat after him that had his same love of life which included, in the goat's case, eating hats.]

Old Hat

Old, dull stuff; out of fashion. [This seems to come from the fact that hat fashions are constantly changing. The fact of the matter is that hat fashions had not been changing very fast at all until the turn of the 19th Century. The expression therefore is likely about 100 years old.]

Mad As A Hatter

Totally demented, crazy. [Hatters did, indeed, go mad. They inhaled fumes from the mercury that was part of the process of making felt hats. Not recognizing the violent twitching and derangement as symptoms of a brain disorder, people made fun of affected hat-makers, often treating them as drunkards. In the U.S., the condition was called the "Danbury shakes." (Danbury, Connecticut, was a hat-making center.) Mercury is no longer used in the felting process: hat-making–and hat-makers–are safe.]

Hat In Hand

A demonstration of humility. For example, "I come hat in hand" means that I come in deference or in weakness. [I assume that the origins are from feudal times when serfs or any lower members of feudal society were required to take off their hats in the presence of the lord or monarch (recall the Dr. Seuss book, *The 500 Hats of Bartholomew Cubbins*). A hat is your most prideful adornment.]

Pass The Hat

Literally to pass a man's hat among members of an audience or group as a means for collecting money. Also to beg or ask for charity. [The origin is self-evident as a man's hat turned upside down makes a fine container.]

HOLD ON TO YOUR HAT

Tight As Dick's Hat Band

Anything that is too tight. [The Dick in this case is Richard Cromwell, the son of England's 17th Century "dictator," Oliver Cromwell. Richard succeeded his dad and wanted to be king but was quickly disposed. The hatband in the phrase refers to the crown he never got to wear.]

Hat Trick

Three consecutive successes in a game or another endeavor. For example, taking three wickets with three successive pitches by a bowler in a game of cricket, three goals or points won by a player in a game of soccer or ice hockey, etc. [From cricket, from the former practice of awarding a hat to a bowler who dismissed three batsmen with three successive balls.]

Hard Hats

In the 19th Century, men who wore derby hats. Specifically Eastern businessmen and later crooks, gamblers and detectives. [Derby hats, a.k.a. Bowlers or Cokes, were initially very hard as they were developed in 1850 for use by a game warden or horseback rider wanting protection.] Today, "Hard Hats" are construction workers [for obvious reasons].

In One's Hat, or In Hat

An expression of incredulity. [Origin unknown.]

Throwing A Hat In the Ring

Entering a contest or a race e.g. a political run for office. [A customer wrote us with the following: "I read in "The Language of American Politics" by William F. Buckley Jr. that the phrase "throw one's hat in the ring" comes from a practice of

19th Century saloonkeepers putting a boxing ring in the middle of the barroom so that customers who wanted to fight each other would have a place to do so without starting a donnybrook. If a man wanted to indicate that he would fight anybody, he would throw his hat in the ring.] At one point, Theodore Roosevelt declared he was running for office with a speech that included a line that went something like, "My hat is in the ring and I am stripped to the waist." The phrase "my hat in the ring" stuck, probably because "I am stripped to the waist" is a little gross.

Hats Off

"Hats off to the U.S. Winter Olympic Team" for example. An exclamation of approval or kudos. [Origins must be from the fact that taking one's hat off or tipping one's hat is a traditional demonstration of respect.]

A Feather In Your Cap

A special achievement. [I assume that the origins on this expression hail from the days when, in fact, a feather for one's cap would be awarded for an accomplishment much like a medal is awarded today and pinned to one's uniform. A feather, or a pin, add a certain prestige or luster to one's apparel.]

Hold On To Your Hat(s)

A warning that some excitement or danger is imminent. [When riding horseback or in an open-air early automobile, the exclamation "hold on to your hat" when the horse broke into a gallop or the car took off was certainly literal.]

Bee In Your Bonnet

An indication of agitation or an idea that you can't let go of and just have to express. [A real bee in one's bonnet certainly precipitates expression.]

HOLD ON TO YOUR HAT

Wearing Many Hats

This of course is a metaphor for having many different duties or jobs. [Historically, hats have often been an integral, even necessary, part of a working uniform. A miner, welder, construction worker, undertaker, white-collar worker or banker before the 1960s, chef, farmer, etc. all wear, or wore, a particular hat. Wearing "many hats" or "many different hats" simply means that one has different duties or jobs.]

All Hat and No Cattle

All show and no substance. For example, in October 2003, Senator Robert Byrd declared that the Bush administration's declarations that it wanted the United Nations as a partner in transforming Iraq were "All Hat and No Cattle". [This Texas expression refers to men who dress the part of powerful cattlemen, but don't have the herds back home.]

To Hang Your Hat (or not)

To commit to something (or not), or stake your reputation on something (or not), like an idea or policy. For example, "I wouldn't hang my hat on George Steinbrenner's decision to fire his manager." [Origin unknown. Can anyone help with this one?]

At the Drop of a Hat

Fast. [Dropping a hat, can be a way in which a race can start (instead of a starting gun for example). Also, a hat is an apparel item that can easily become dislodged from its wearer. Anyone who wears hats regularly has experienced the quickness by which a hat can fly off your head.]

Interviews with Bern Porter

To Tip Your Hat or A Tip of the Hat

An endorsement of respect, approval, appreciation, or the like. Example: "A tip of the hat to American troops for the capture of Saddam Hussein." [This is simply verbalizing an example of hat etiquette. Men would (and some still do) tip their hat to convey the same message.]

My Hat Instead of Myself

This is an expression from Ecuador, home of the "Panama" hat. It means what is says; it is preferable to give up your hat than your life. [The Guayas River runs through Guayaquil, Ecuador's largest city on the Pacific coast. People from the city were known to hunt alligators for their hides in the river by swimming stark naked wearing Panama hats on their heads and long knives between their teeth. When the reptiles open their jaws and go for the swimmer, he dives leaving his hat floating on the surface for the alligator to chew on while he plunges the knife into the animal's vitals. From *The Panama Hat Trail* by Tom Miller.]

Bad Hat

I believe this is a French expression for a bad person. [Ludwig Bemelmans' *Madeline* series of children's books, set in France, includes one *Madeline and the Bad Hat*. In this story Madeline, our heroine, refers to a little boy neighbor as a "bad hat." She clearly means this as a metaphor for a bad person and because I do not know the expression in English, I assume this is a common French reference.]

Hat by Hat

Step by step. Nevada Barr's book *Seeking Enlightenment: Hat by Hat* means just that.

HOLD ON TO YOUR HAT

Keeping Something Under One's Hat

Keeping a secret. [People kept important papers and small treasures under their hats. One's hat was often the first thing put on in the morning and the last thing taken off at night, so literally keeping things under one's hat was safe keeping. A famous practitioner of this was Abraham Lincoln. The very utilitarian cowboy hat was also commonly used for storage.]

Here's Your Hat, But What's Your Hurry

When someone has taken up enough of your time and you want him/her to leave. [Origin unknown.]

Carry His Office in His Hat

Operating a business on a shoestring. [Important papers and the like were often carried in one's hat.]

Sets Her Cap

A young lady "sets her cap" for a young man who she hopes to interest in marrying her. [Long ago, maidens wore caps indoors because homes were poorly heated. A girl set her most becoming hat on her head when an eligible fellow came to call.]

Thinking Cap

To put on your "thinking cap" is to give some problem careful thought. [Teachers and philosophers in the Middle Ages often wore distinctive caps that set them apart from those who had less learning. Caps became regarded as a symbol of education. People put them on (literally or figuratively) to solve their own problems.]

Black Hat

Black hat tactics, black hat intentions, etc. refer to nefarious actions or designs. [Black hats in Western lore and literature were the bad guys.]

White Hat

Although I don't see or hear this expression as much as "Black Hat," it simply is the opposite of the above. [Good guys wore/wear white hats.]

Both Black Hat and White Hat have new meanings in the context of Internet/IT hacking.

Go S**t In Your Hat

An insult to someone who expects something unreasonable from you, e.g. "No, you can't have my parking space, Buddy. Why don't you go s**t in your hat and pull it down over your ears." [An old-school expression with roots from the WW II era.]

Bibliography

BOOKS BY PORTER

Doldrums: A Study in Surrealism. A.I. Press. Newark, New Jersey. 1941.

Water Fight. First edition: A.I. Press. Newark, New Jersey. 1941; Second editions: L.A.
Press. Pine Hill Printery, South Dakota and Washington. 1941.

Art Techniques. Gillick Press. Berkeley, California. 1947.

Drawings: 1955-56. Bern Porter Editions. Berkeley, California. 1957.

Physics for Tomorrow. Rockcliff Printers. Burnie, Tasmania, 1959.

Aphasia: A Psycho-Visual Satire. Bern Porter Editions, Madison, Maine. 1961.

Scandinavian Summer: A Psycho-Visual Recollection. Bern Porter Editions. Madison, Maine. 1961.

What Henry Miller Said and Why It Is Important. Introduction by John G. Moore. Marathon Press. Pasadena, California. 1964.

Art Productions: 1928-1965. Marathon Press. Pasadena, California. 1965.

468B: Thy Future. Huntsville, Alabama. 1966.

Dieresis. Bern Porter Editions. Rockland, Maine. 1969.

I've Left. First edition: Marathon Press. Pasadena, California. 1963; Second edition: Introduction by Dick Higgins. Something Else Press. Millerton, New York. 1971.

Knox County Maine, A Regional Report, 1969 Summary. Knox County Regional Planning Commission, 1969.

Found Poems. Something Else Press. Millerton, New York. 1972.

The Manhattan Telephone Book. First editions: Abyss Publications, Somerville, Massachusetts. 1972; Second Edition: Dylux Edition, 1975.

The Wastemaker: 1926-1961. Introduction by Richard Kostelanetz. Abyss Publications. Somerville, Massachusetts. 1972.

Where to go what to do when in New York, week of June 17, 1972. Bern Porter Editions. 1974.

Run-On. Bern Porter Editions. Belfast, Maine. 1975.

Selected Founds. Croissant and Company. Athens, Ohio. 1975.

Interviews with Bern Porter

Gee-Whizzels. Maine Coast Printers. Rockland, Maine. 1977.

American Strange. Special edition of Spanner (#14). London, England. 1978.

Isla Vista. Turkey Press. Isla Vista, California. 1981.

The Book of Do's. The Dog Ear Press / Tilbury House Publishers. Gardiner, Maine. 1982.

Here Comes Everybody's Don't Book. The Dog Ear Press / Tilbury House Publishers. Gardiner, Maine, 1984.

Horizontal Hold. w/ Todd Transformer. A.P.S.U. Clarksville, Tennessee. 1985.

The Last Acts of Saint Fuck You. Xexoxial Editions. Madison, Wisconsin. 1985.

My My Dear Me. Xexoxial Editions. Madison, Wisconsin. 1985.

Neverends. The Runaway Spoon Press. Port Charlotte, Florida. 1988.

Why My Left Leg is Hot. Xexoxial Editions. Madison, Wisconsin. 1988.

Sweet End. The Dog Ear Press / Tilbury House Publishers. Gardiner, Maine. 1989.

Vocrescends. w/ Malok. The Runaway Spoon Press. Port Charlotte, Florida. 1990.

CRCNCL / A sur surrealistic video script. Xexoxial Editions. Madison, Wisconsin. 1991.

Frozen Hypnosis (#1-29). w/ Malok. Waukau, Maine. 1991 - ?.

Numbers. Introduction by Erika Pfander. The Runaway Spoon Press. Port Charlotte, Florida. 1991.

Untitled. Special edition of Xerolage (#16). Xexoxial Editions. Madison, Wisconsin. 1991.

Less Than Overweight. Plaster Cramp Press. Chicago, Illinois. 1992.

Sounds That Arouse Me. Ed. and with an introduction by Mark Melnicove. Tilbury House Publishers. Gardiner, Maine. 1992.

Book of Wisdom. 1995.

Symbols. Introduction by Erika Pfander. The Runaway Spoon Press. Port Charlotte, Florida. 1995.

Henry Miller's Semblance of a Devoted Past: A Study in Censorship. By Roger Jackson. Afterword by Bern Porter. Roger Jackson, Publisher. Ann Arbor, Michigan. 1995.

Questions About Henry Miller That No One Ever Asked Me - With Answers. Roger Jackson Publishers. Ann Arbor, Michigan. 1995.

HOLD ON TO YOUR HAT

These 50 Years Gone. Roger Jackson, Publisher. Ann Arbor, Michigan. 1995.

My Affair with Anaïs Nin (Part I: Paris-New York Days). w/ Natasha Bernstein. Roger Jackson Publishers. Ann Arbor, Michigan. 1996.

Bern Porter's Pillow Book. Roger Jackson, Publisher. Ann Arbor, Michigan. 1996.

The Best Period of My Life. Roger Jackson, Publisher. Ann Arbor, Michigan. 1996.

Night Thoughts on Henry Miller, Ben Abramson and Claude Houghton. Roger Jackson Publisher. Ann Arbor, Michigan. 1996.

A Sex Oriented, Woman Connected Guy Doing His Own Thing: Bern Porter on Henry Miller, A Manuscript. Roger Jackson, Publisher. Ann Arbor, Michigan. 1996.

Signs. TheRunaway Spoon Press. Port Charlotte, Florida. 1996.

These 50 Years Gone along with The Sorrow. Roger Jackson, Publisher. Ann Arbor, Michigan. 1996.

My Affair with Anaïs Nin (Part II: San Francisco Days). Roger Jackson, Publisher. Ann Arbor, Michigan. 1997.

My Affair with Anaïs Nin (Volume III: Berkley Days). w/ Natasha Bernstein. Illustrated by Al Berlinski. Roger Jackson, Publisher. Ann Arbor, Michigan. 1997.

Crossfire. w/ Jack Saunders. Roger Jackson, Publisher. Ann Arbor, Michigan. 1997.

I Pursue Her Still: Bern Porter on Anaïs Nin. Roger Jackson, Publisher. Ann Arbor, Michigan. 1997.

Lava Love: A Postcard Exchange. w/ Marilyn "Mo" Monroe. Roger Jackson, Publisher. Ann Arbor, Michigan. 1997.

Open Letter to O. J. Simpson. Roger Jackson, Publisher. 1997.

Rites of Spring. Roger Jackson, Publisher. Ann Arbor, Michigan. 1997.

Redundant, Redundant. Roger Jackson, Publishers. Ann Arbor, Michigan. 1997.

A Walk on the Wild Side: A Photo Tour of the Sculpture Garden at the Institute of Advanced Thinking. 23-item card and photo set. Roger Jackson, Publisher. Ann Arbor, Michigan. 1997.

My Affair with Anaïs Nin (Part IV: Silver Lake Days). w/ Natasha Bernstein. Roger Jackson, Publisher. Ann Arbor, Michigan. 1998.

As Birds Fly: A Drama in Five Scenes. w/ Mary Weaver. Roger Jackson, Publisher. Ann Arbor, Michigan. 1998.

Interviews with Bern Porter

Bern Porter's Book of Wisdom, Roger Jackson, Publisher. Ann Arbor, Michigan. 1998.

Founds (#1-9). X-Ray Book Co. San Francisco, California, 1998.

Monica Lewinsky, We All Want Some of Yours. Roger Jackson, Publisher. Ann Arbor, Michigan. 1998.

FROM: Bern Porter TO: The World! Roger Jackson, Publisher. Ann Arbor, Michigan. 1999.

That Trio Again: Monica, Hillary, and Bill. Illustrated by Al Berlinski. Roger Jackson, Publisher. 1999.

Bern Porter's Book of Light: A Male Art Documentation. 1999.

Physics Today. Roger Jackson, Publisher. Ann Arbor, Michigan. 1999. [Slightly edited reprint of *Physics for Tomorrow*, 1959.]

Monica, Monica. Illustrated by Gene King. Roger Jackson, Publisher. Belfast Common. 1999.

Let Me Do Yours Monica. 1999.

Insider's Guide to Belfast Maine. 2000.

Bern on Bern. 2000.

So Far: A Bern Porter Miscellany. Roger Jackson, Publisher. Ann Arbor, Michigan. 2001.

Wuondruskh, A 2004 Bookdust Resurrection, Xexoxial Editions, West Lima, Wisconsin. 2004.

Infreucombia, Xexoxial Editions, West Lima, Wisconsin, 2008.

Where To Go, What To Do, When You Are Bern Porter: Week Of June 17, 1972. Complete reprint and faithful facsimile of original published by Porter in 1974, with essay by Mark Melnicove. Esopus, Issue #12. 2009.

Found Poems, Nightboat Books. Reprint of 1972 Something Else Press edition, with Preface by David Byrne, Introduction by Joel Lipman, Afterword by Mark Melnicove. 2011.

Tapes, Films and Videos

Wildcat Hill Revisited. Film study of a day in the life of Edward Weston. With Harry Bowden. 1947.

HOLD ON TO YOUR HAT

Black Box #4. American magazine of recorded poetry. 12'14" of sound poetry. Washington, D.C., 1975.

Earum Magnus. Super-8 film conceived by Bern Porter. Music by Charlie Morrow. Directed by Dick Higgins. Photography and sound by Scott B. 22'. 1979.

The Eternal Poetry Festival. Sound poetry improvisations with Mark Melnicove. Xexoxial Editions. Madison, Wisconsin. 1979.

Found Sounds. New Wilderness Audiographics. New York, 1981. Reissued by Xexoxial Editions. Madison, Wisconsin. Side one is Porter with Dick Higgins and Charlie Morrow recorded in 1978; side two is Porter in concert with Patricia Burgess, Charlie Morrow, and Glen Velez.

Aspects of Modern Poetry. Porter talking with Robert Holman. Recorded from broadcast on WBAI, New York City. 1982.

Jacqualin and Rosa. Found sound radio play by Porter. New Wilderness Audiographics. *Found Sounds* #7913A, side 1, section 3, New York, 1982

Why My Left Leg is Hot. Book and video, Xexoxial Editions, Madison, Wisconsin, 1987.

Williamson Street Blues. Includes readings of *Last Acts*, Chinese poetry, and the Madison phone book; also a rock version of *Last Acts.* Xexoxial Editions. Madison, Wisconsin. 1988.

Williamson Street Night. Xexoxial Editions. Madison, Wisconsin. 1989. Side one is Porter reading from Abraham Lincoln Gillespie and mIEKAL aND from his own book, RAW. SWAY. ALOUD., accompanied by a recording of the Abenaki Indians. Side two is Porter reading from Abraham Lincoln Gillespie and Elizabeth Was reading ROOMS by Gertrude Stein.

December Ninth. w/ Jo Brewer and Elizabeth Dunker. A Jeff Brewer recording. Newton Center, Massachusetts. 1989.

Sleepers Awake. Performance from Kenneth Patchen's book by Porter and Mark Melnicove at the Second Kenneth Patchen Festival. Warren, Ohio. Published by Cedar Grove House. Dresden, Maine. 1990.

Belfast Berning. A video of Porter's 80th birthday celebration. VHS. 22'. Produced by Vanessa Barth. Cyclops Video. Freeport, Maine. 1991.

Interviews with Bern Porter

Why My Left Leg is Hot. DVD. 1 minute and 1 hour versions. Re-edit by Camille Bacos, mIEKAL aND, and jUStin katKO. 2005.

CRCNCL. DVD. 1 hour. Re-edit by Camille Bacos. 2005.

COLLECTIONS

Special Collections, University of California at Los Angeles Library. This is the major collection of Porter's work, including material as yet un-catalogued. The collection includes original books by Porter as well as magazines and rare "one-of-a-kind" items created by Porter, such as "Hand Painted Chocolates" (containing Sciart fortunes on slips of paper), "14th of February" (in a heart-shaped slipcase prepared by Porter and remodelled from a box of chocolates), and "Hexa 914" (in a six-sided box with cellophane cover revealing page after page of nine circles in bright blue, purple, and orange then repeated in different variations).

Special Collections, John Hay Library, Brown University, Providence, Rhode Island. This collection includes original books by Porter as well as books published by him, a copy of his master plan for Knox County, Maine, several of his scientific articles, and his master's thesis in physics, and correspondence and materials gathered by biographer James Schevill.

Special Collections, Colby College Library, Waterville, Maine. In addition to original books by Porter and books published by him, this collection contains letters to Porter, a complete set of CIRCLE magazines, and many books with inscriptions by the authors to Porter – a valuable record of Porter's reading interests. Also collected are photographs, letters, and manuscripts, including material relating to his various world cruises.

Special Collections, Bowdoin College Library, Brunswick, Maine. This collection includes copies of Porter's own books and rare copies of books that he published; also photographs, letters, and manuscripts, including material relating to his various world cruises.

Bern Porter Archive in the library of the Museum of Modern Art, New York City. This collection includes, in addition to rare copies of Porter's books and books published by him, a great deal of material about Porter's involvement with Mail Art.

DECOR by BERN PORTER
ALL RIGHTS WHATEVER RESERVED

Being a Map of Physics

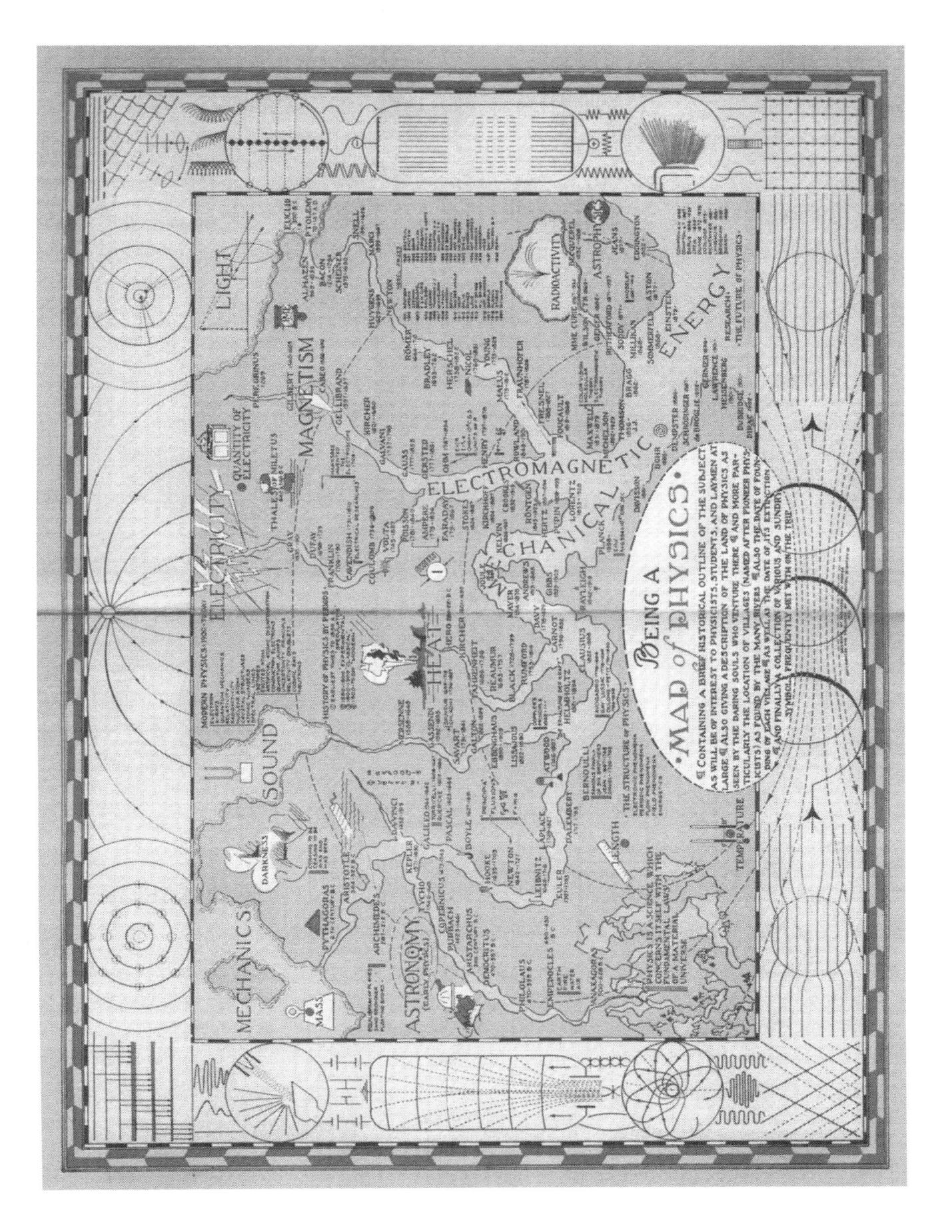

¶ Containing a brief historical outline of the subject, as will be of interest to physicists, students, and laymen at large. ¶ Also giving a description of the Land of Physics as seen by the daring souls who venture there. ¶ And more particularly the location of villages (named after pioneer physicists) as found by the many rivers. ¶ Also the date of founding by each village, as well as the date of its extinction. ¶ And finally a collection of various and sundry symbols frequently met with on the trip.

JOYCEANA

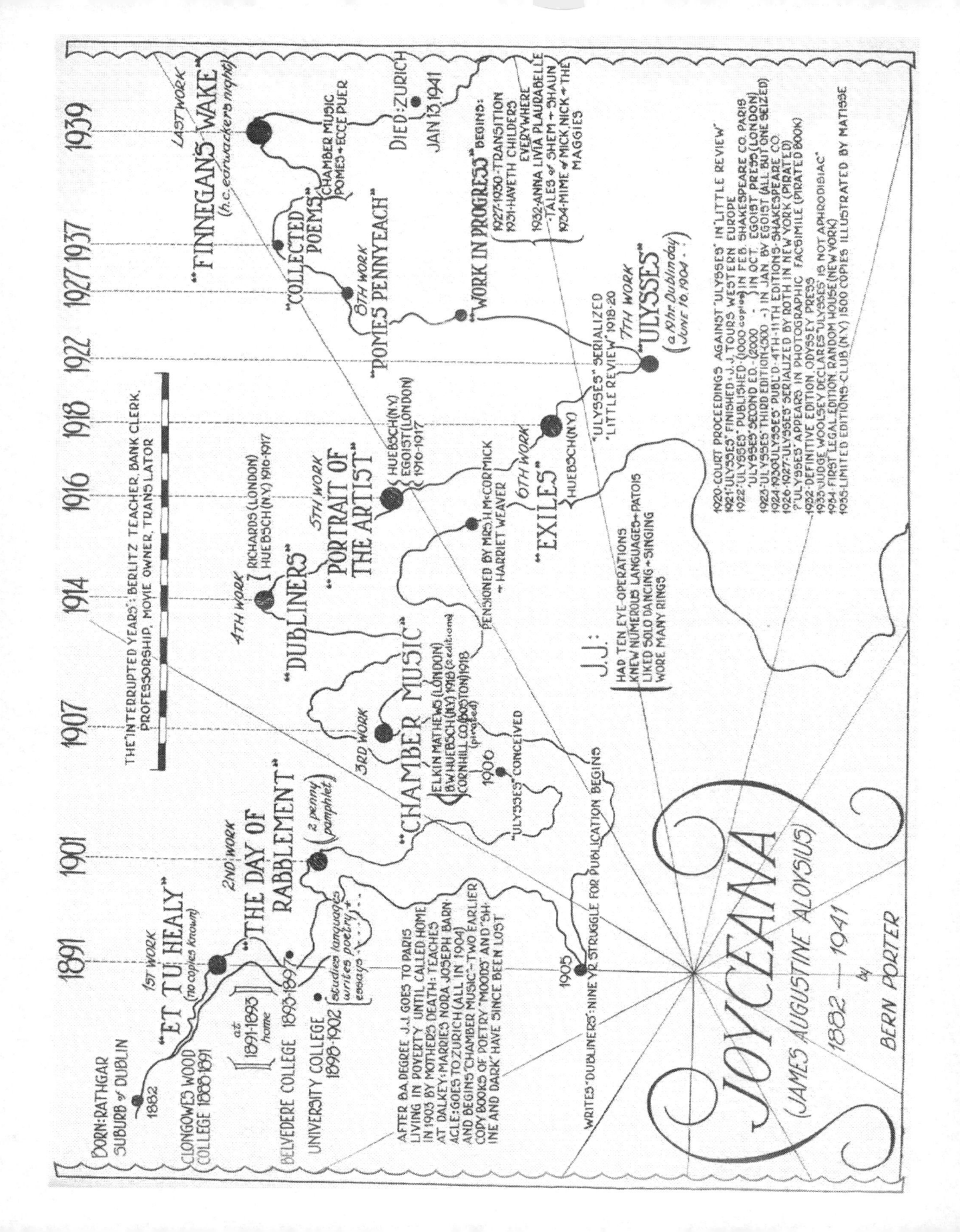

(JAMES AUGUSTINE ALOYSIUS)

1882 — 1941

by BERN PORTER

1882 BORN: RATHGAR SUBURB OF DUBLIN
CLONGOWES WOOD COLLEGE 1888-1891
BELVEDERE COLLEGE 1893-1897
UNIVERSITY COLLEGE 1898-1902 — studies languages, writes poetry + essays

1891 "ET TU, HEALY" 1ST WORK (no copies known)
at 1891-1893 home

1901 "THE DAY OF RABBLEMENT" 2ND WORK (2 penny pamphlet)

AFTER BA DEGREE J.J. GOES TO PARIS LIVING IN POVERTY UNTIL CALLED HOME IN 1903 BY MOTHER'S DEATH; TEACHES AT DALKEY; MARRIES NORA JOSEPH BARNACLE; GOES TO ZURICH (ALL IN 1904) AND BEGINS "CHAMBER MUSIC" — TWO EARLIER COPY BOOKS OF POETRY "MOODS" AND "SHINE AND DARK" HAVE SINCE BEEN LOST

1905 WRITES "DUBLINERS"; NINE YR. STRUGGLE FOR PUBLICATION BEGINS

1906 "ULYSSES" CONCEIVED

1907 "CHAMBER MUSIC" 3RD WORK
{ELKIN MATHEWS (LONDON)
(B.W. HUEBSCH (N.Y.) 1918 (2 editions)
CORNHILL CO. (BOSTON) 1913 (pirated)}

1914 "DUBLINERS" 4TH WORK
{ RICHARDS (LONDON)
HUEBSCH (N.Y.) 1916-1917 }

1916 "PORTRAIT OF THE ARTIST" 5TH WORK
{ HUEBSCH (N.Y.)
EGOIST (LONDON) 1916-1917 }

THE "INTERRUPTED YEARS": BERLITZ TEACHER, BANK CLERK, PROFESSORSHIP, MOVIE OWNER, TRANSLATOR

PENSIONED BY MRS. H. McCORMICK + HARRIET WEAVER

1918 "EXILES" 6TH WORK HUEBSCH (N.Y.)

"ULYSSES" SERIALIZED "LITTLE REVIEW" 1918-20

1922 "ULYSSES" 7TH WORK (a 12 hr. Dublinday) (June 16, 1904 ...)

1927 "WORK IN PROGRESS" BEGINS:
1927-1930 - TRANSITION
1931 - HAVETH CHILDERS EVERYWHERE
1932 - ANNA LIVIA PLAURABELLE
"TALES of SHEM + SHAUN"
1934 - MIME of MICK NICK + THE MAGGIES

1937 "COLLECTED POEMS" 8TH WORK
CHAMBER MUSIC
POMES + ECCE PUER

"POMES PENNYEACH"

1939 "FINNEGAN'S WAKE" LAST WORK (h.c. earwickers night)

DIED: ZURICH JAN 13, 1941

J.J.:
HAD TEN EYE OPERATIONS
KNEW NUMEROUS LANGUAGES + PATOIS
LIKED SOLO DANCING + SINGING
WORE MANY RINGS

1920 - COURT PROCEEDINGS AGAINST "ULYSSES" IN "LITTLE REVIEW"
1921 - "ULYSSES" FINISHED; J.J. TOURS WESTERN EUROPE
1922 - "ULYSSES" PUBLISHED IN FEB. SHAKESPEARE CO. PARIS
"ULYSSES" SECOND ED. — IN OCT. EGOIST PRESS (LONDON) (1000 copies)
1923 - "ULYSSES" THIRD EDITION — 500 — IN JAN. BY EGOIST (ALL BUT ONE SEIZED)
1924 - 1930 "ULYSSES" PUBL'D. 4TH - 11TH EDITIONS SHAKESPEARE CO.
1926 - 1927 "ULYSSES" SERIALIZED BY ROTH IN NEW YORK (PIRATED)
? "ULYSSES" APPEARS IN PHOTOGRAPHIC FACSIMILE (PIRATED BOOK)
1932 - DEFINITIVE EDITION ODYSSEY PRESS
1933 - JUDGE WOOLSEY DECLARES "ULYSSES" IS NOT APHRODISIAC
1934 - FIRST LEGAL EDITION RANDOM HOUSE (NEW YORK)
1935 - LIMITED EDITIONS-CLUB (N.Y.) 1500 COPIES ILLUSTRATED BY MATISSE

Made in the USA
Charleston, SC
13 March 2017